8051 Microcontroller Architecture, Programming and Application

8051 Microcontroller Architecture, Programming and Application

M. Mahalakshmi
M.Sc.,M.Phil
Department of Computer Science
Sri Vasavi College,
Erode, Tamilnadu

UNIVERSITY SCIENCE PRESS
(An Imprint of Laxmi Publications Pvt. Ltd.)
An ISO 9001:2015 Company
BENGALURU • CHENNAI • GUWAHATI • HYDERABAD
KOCHI • KOLKATA • LUCKNOW • RANCHI
NEW DELHI

Printed and bound in India
First Edition : 2012; *Edition* : 2025
ISBN : 978-93-81159-24-8

Branches

✆	Bengaluru	080-26 75 69 30
✆	Chennai	044-24 34 47 26
✆	Guwahati	0361-254 36 69
✆	Hyderabad	040-27 55 53 83
✆	Kochi	0484-405 13 03
✆	Kolkata	033-40 04 77 79
✆	Lucknow	0522-430 36 13
✆	Ranchi	0651-224 24 64

PUBLISHED IN INDIA BY

UNIVERSITY SCIENCE PRESS
(An Imprint of Laxmi Publications Pvt. Ltd.)

An ISO 9001:2015 Company
113, GOLDEN HOUSE, GURUDWARA ROAD, DARYAGANJ,
NEW DELHI - 110002, INDIA

Corporate Office : B-33, Sector-5, Noida,
Gautambuddha Nagar (U.P.) - 201 301, INDIA
Telephone : +91-120-44 46 900
www.laxmipublications.com
info@laxmipublications.com

FOREWORD

It's my pleasant privilege to write a foreword for this book "8051 Microcontroller Architecture, Programming and Application". The purpose of this book is to present, as clearly as possible, the principles of 8051 microcontroller.

The contents of the book are well organized and written in a simple language with numerous worked example and exercise problems with answers. The level of presentation is suitable for self study. The book is intended for introductory microcontroller courses at the under graduate level in technology and engineering.

Myself being a teacher in the field of computer science. I strongly recommend this book to every electronics and computer science students. The questions at the end of the chapters not only clearly underline the author's experience and sound theoretical knowledge, but also indicate her deep commitment to the subject. This book will be boon to the microcontroller beginners and experts.

Prof. B.Mahalingam,
Head, Department of Computer Science,
Sri Vasavi College,
Vasavi College PO,
Erode – 638 316.

PREFACE

It is written for the individual who wishes to learn the Microcontroller. The material in this book is appropriate for an introductory course in digital logic in either a computer or an electronic program. It is also appropriate for the self study and as reference for individuals working in this field.

At the end of each chapter have review questions, called self tests, which are intended to be a self check of key ideas and concepts. In any subject area, there are many terms and concepts to be learned.

Every major concept is worked out through examples, to a numerical conclusion. The student can work any example to test the validity of the concept and draw their own conclusions. In addition to the numerous examples, each chapter concludes with a wide variety of questions.

This book aims at providing the students with the understanding of the basic operating principles of microcontroller. A text should be sufficiently clear to enable the reader to understand the material well by its reading with realism.

All topics have been explained in simple language with illustrations, block diagrams, specifications, applications, comparative table and examples. Summary has been added at the end of the each chapter and besides the review questions, the objective type questions and numerical questions have been given to help the students.

I hope the book will be found useful by the students and instructors alike. I have benefited from the assistance of a number of people in reviewing, writing and producing this text. I would like to take this opportunity to thank the following reviewers who providing many helpful, construction and suggestions.

—Author

CONTENTS

Chapter 1

Microprocessor and Microcontroller

1.1 INTRODUCTION

The integrated circuit from an Intel 8742, an 8-bit micro controller that includes a CPU running at 12 MHz, 128 bytes of RAM, 2048 bytes of EPROM, and I/O in the same chip.

A microcontroller (also mcu or μc) is a computer-on-a chip. It is a type of microprocessor emphasizing high integration, low power consumption, self-sufficiency and cost- effectiveness, in contrast to a general-purpose microprocessor (the kind used in a pc). In addition to the usual arithmetic and logic elements of a general purpose microprocessor, the micro controller typically integrates additional elements such as read-write memory for data storage, read-only memory such as flash for code storage, EEPROM for permanent data storage, peripheral devices and input/output interfaces. At clock speeds of as little as a few MHz or even lower, microcontroller often operate at very low speed compared to modern day microprocessors, but this is adequate for typical application. They consume relatively little power (milliwatts), and will generally have the ability to sleep while waiting for an interesting peripheral event such as button press to wake them up again to do power consumption while sleeping may be just nano watts, making them ideal for low power and long lasting battery applications.

1.2 MICROPROCESSOR AND MICROCONTROLLER

1.2.1 Microprocessor

It is the heart of the microcontroller system. It consists of Arithmetic Logic Unit (ALU), registers and control circuit. The arithmetic and logic operations are carried out by the ALU. The microprocessor executes the program stored in the memory in a sequence.

The microprocessor is a semiconductor device manufactured the VLSI techniques.

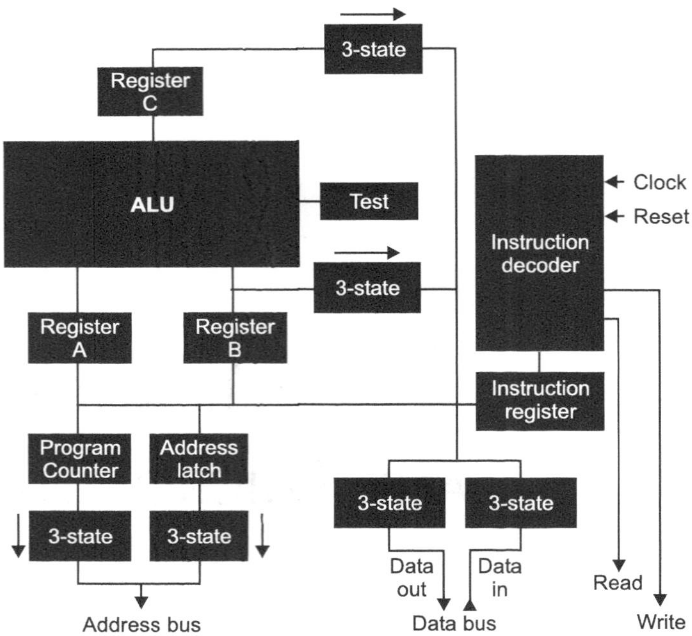

Fig. 1.1 Microprocessor Architecture

This is about as simple as a microprocessor gets. This microprocessor has:

- An **address bus** (that may be 8, 16 or 32 bits wide) that sends an address to memory
- A **data bus** (that may be 8, 16 or 32 bits wide) that can send data to memory or receive data from memory
- An **RD** (read) and **WR** (write) line to tell the memory whether it wants to set or get the addressed location
- A **clock line** that lets a clock pulse sequence the processor
- A **reset line** that resets the program counter to zero (or whatever) and restarts execution

1.2.2 Microcontroller

The Intel 8051 is single chip microcomputer which was developed by Intel in 1980 for use embedded system. It was popular in the 1980's and the early 1990's, but today it has largely been superseded.

Intel's original 8051 family was developed using NMOS technology, but later versions identified by a letter "c" in their name. Example 80C51 used CMOS technology and were less power hungry than their NMOS predecessors-this made them eminently more suitable for battery power devices.

A particularly useful features of the 8051 core is the inclusion of a Boolean processing engine which allows bit level Boolean logic operation to be carried out directly and efficiently on internal register and RAM. The features of boolean processing helped to cement the 8051's popularity in industrial control applications.

Another features is that it has form separate register sets, which can be used to greatly reduce interrupt context in a stack.

Microcontrollers are frequently used in automatically controlled products and devices, such as automobile engine control systems, remote controls, office machines, appliances, power tools and toys. By reducing the size, cost and power consumption compared to a design using a separate microprocessor, memory and input/output devices, microcontrollers make it economically to electronically control many more process.

Microcontrollers are hidden inside a surprising number of products these days. If your micro wave oven has LED or LCD screenland a keypad, it contains a microcontroller and can have as many as six or seven. The engine is controlled by a microcontroller as are the anti locks brakes, the cruise control and so on. Any device that has a remote control almost certainly contains a microcontroller. TVs, UCRs, answering machines, laser printers, telephones (the ones with caller id, 20 - number memory, etc.) pagers and feature-laden refrigerators, dishwashers, washers and dryers (the ones with display and keypads) you get the idea, basically, any product or device that interacts with its users has a microcontroller buried inside.

Microcontrollers are "embedded" inside some devices (often a consumer product) so that they can control the features or actions of the product. Another name for a microcontroller, therefore is "embedded controller".

Microcontrollers are dedicated to one task and run one specific program. Program is stored in ROM (read only memory) and generally does not change.

Microcontrollers are often lower-power devices. A desktop computer is almost plugged into a wall socket and might consume 50 watts of electricity. A battery operated microcontroller might consume 50 milliwatts.

A microcontroller has a dedicated input device and often (but not always) has a LED or LCD display for output. A microcontroller also takes input from the devices it is controlling and controls the device by sending signals to different components in the device. For example: the microcontroller controls the channel selector, the speaker system and certain adjustments on the picture tube electronics such as tint and brightness. The engine controller in a car takes input from sensors such as the oxygen and knock sensors and controls things like fuel mix and spark plug timing. A microwave oven controller takes input from a keypad, displays output on an LCD display and controls the relays that turns the microwave generator on and off. A microcontroller is often small and low cost components are chosen to minimize size and to be as inexpensive as possible.

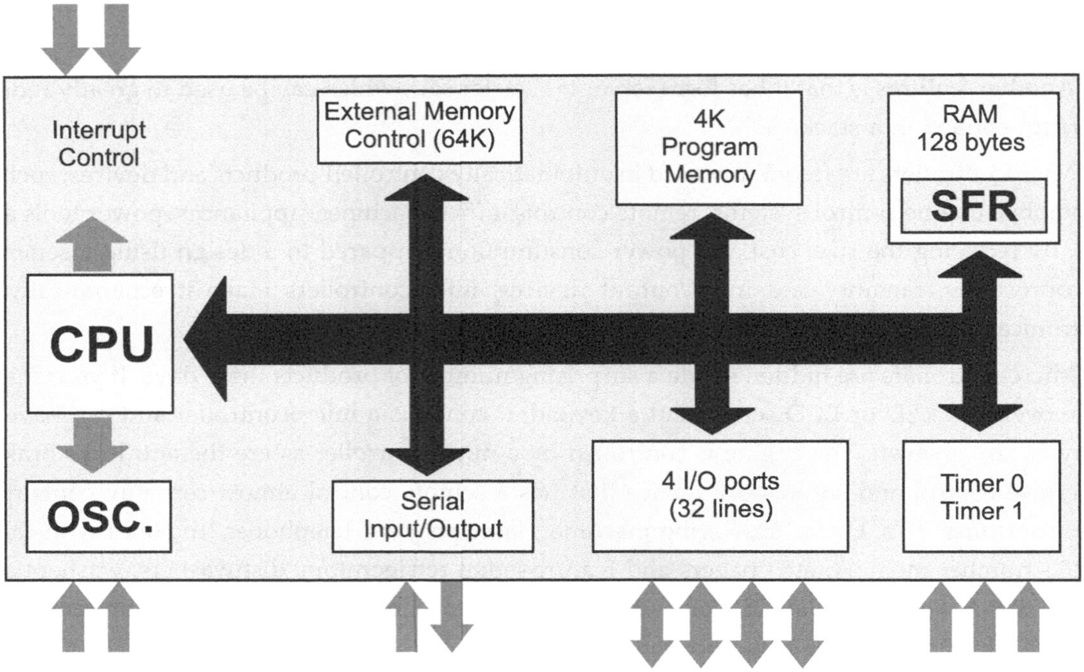

Fig. 1.2 Microcontroller Model

1.2.3 Difference Between Microprocessor and Microcontroller

Microprocessor	Microcontroller
1. Microprocessor contains ALU, general-purpose register, stack, pointer, program counter, clock timing circuit and interrupts circuit.	1. Microcontroller the circuitry of microprocessor and in addition it has built in ROM, RAM, input/output devices timers and counters.
2. It has many operational codes (op codes) for moving only data from external memory.	2. It has one or two op codes.
3. It has one or two bits handling instruction.	3. It has many bits handling instruction.
4. Time taken to complete a process is more.	4. Time taken to complete a process is less.
5. Microprocessor based system requires more Hardware.	5. Microcontroller based systems requires less Hardware reducing PCB size and increasing the reliability.
6. Microprocessor based system is more flexible in designing point of view.	6. Microcontroller based system is less flexible in designing point of view.
7. Not capable of handling Boolean function.	7. Capable of handling Boolean functions.
8. Data size in microprocessor varies from 8-bits to 64-bits.	8. Data size in microcontroller varies from 4-bits to 32-bits.
9. Less number of pins are multi functioned.	9. More number of pins are multifunctional.
10. It has single memory map for data and code.	10. It has separate memory map for data and code.

11. It has CPU, memory addressing circuits and interrupt handling circuits. (e.g) 8085, Z80 etc.	11. It has CPU with timer's parallel and serial input/output and internal RAM, ROM. (e.g) MCS51, 8048 etc.

1.3 MICROCOMPUTER ORGANIZATION

1.3.1 Introduction

In a very general sense any microcomputer (example a pc) is best regarded as a system in corporating CPU and associated hardware whose purpose is to manipulate data in some passion. This is exactly what any digital circuit designed using SSIs and MSIs (*i.e.*, gates, flip-flops etc.) will also do.

The microprocessor at the heart of the microcomputer is the best regarded as a general purpose logic device one can say that a microcomputer is a assembly of a devices, including a CPU, which manipulates data depending on one or more inputs and according to a program, in order to generate one or more outputs. With the understanding that one can change the data manipulation being done, merely by changing the program. This is probably a better definition of a microcomputer.

1.3.2 Basic Components of a Microcomputer

The microprocessor is the basic computers of a microcomputer system. The functional blocks shown in Fig. 1.3.

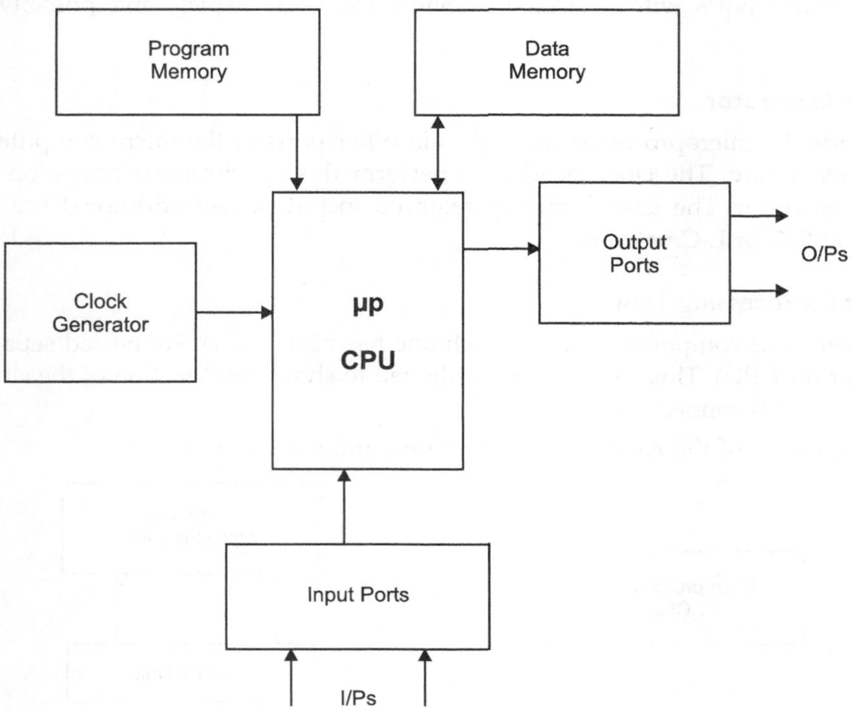

Fig. 1.3 Block Diagram of Microcomputer

1.3.3 Program Memory

The program memory is the microcomputer system component where this instructions sequence is stored. In initialization usually a power up or manual reset – the processor starts by executing the instruction in a predetermined location in program memory are generally finds its part of the program stored in read only memory of ROM, *i.e.*, the program memory is implemented in ROM. Additional program may be located into the data memory by the ROM program.

1.3.4 Data Memory

The functional blocks in the micro computer used for this storage in the data memory. Note that a micro controller will typically have some internal register which can also be used, if available for such storage. The portion of the data memory used as temporary storage, of intermediate results, or the storing and retrieving data must have read and write capability. The data memory segment used for this storage is typically implemented using ROM's, PROM's, or EPROM's.

1.3.5 Input Ports

The input ports allow data to pass from the outside world to the micro computer – data which will be used in the data manipulation being alone by the micro computer. The key sound will interface to this system via input port.

1.3.6 Output Ports

The output ports provide the micro computer the capability to communicate with the outside world. The output ports are used by the micro computer to send data to devices outside the system. The output ports will be used to devices the CRT display, and possibly other LED display.

1.3.7 Clock Generator

Operation inside the microprocessor, as well as in other parts of the micro computer, is usually synchronous by nature. The clock needed to perform this synchronous operation is provided by the clock generator. The user is merely required to put proper additional timing circuitry, example crystal R-C or L-C network.

1.3.8 Central Processing Unit

We have shown these components, where each one has been shown connected separately to the microprocessor (or CPU). This connection is indented to show the direction of the single transfer, rather than the actual connection.

The CPU consists of the Arithmetic/Logical unit and Control Unit.

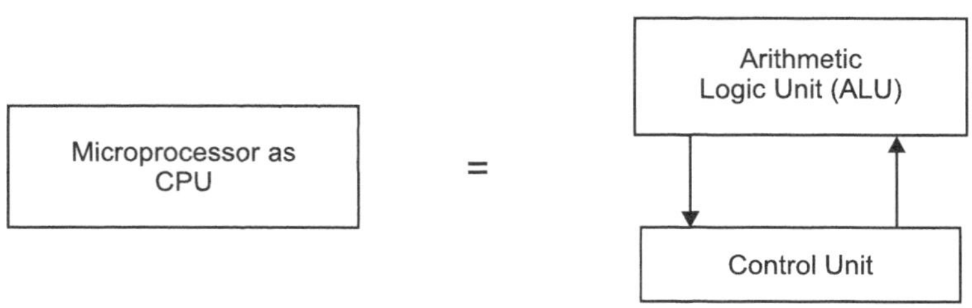

Fig. 1.4 Microprocessor as CPU

The CPU contains various registers to store data, ALU to perform arithmetic and logical operations, instruction decoders, and counters and control lines. The CPU reads instruction from the memory and performs the task specified. It communicates with input / output devices either to accept or to send data. These devices are also known as peripherals. The CPU is the primary and center player in communicating with devices. Such as memory, input and output. However, the timing of the communication process is controlled by the group of circuits called the Control Unit.

1.4 EVOLUTION OF MICROPROCESSOR

The CPU in a single LSI or in a single VLSI is called a microprocessor. A digital computer whose CPU is a microprocessor is called a microcomputer.

First generation microprocessor

The microprocessor was introduced in the year by Intel Corporation, USA. It was a 4 bit microprocessor having 16 pins housed completely in a single chip of PMOS technology. The Intel Corporation released the INTEL8008 version of the microprocessor in the year 1972.

A list of **first generation** microprocessors are given below:

Microprocessor	Word size
* INTEL 4004	4 bit
* INTEL 4040	4 bit
* FAIRCHILD PPS25	4 bit
* NATIONAL IMP4	4 bit
* ROCKWELL PPS4	4 bit
* INTEL 8008	8 bit
* NATIONAL IMP8	8 bit
* ROCKWELL PPS8	8 bit
* MOSTEK 5065	8 bit
* NATIONAL IMP	16bit
* NATIONAL PACE	16 bit

Applications

4 bit microprocessors are used in simple applications such as calculators, electronic toys, video games etc.

Second-generation microprocessors

The second-generation microprocessors were introduced in the year 1973. They were manufactured using NMOS technology. This technology offers faster operation than PMOS.

The INTEL8080 an 8 bit microprocessor designed using NMOS technology. It is much faster and has many more instructions than INTEL8008.

Some of the third generations are given

Microprocessor	Word size
INTEL 8080	8 bit
INTEL 8085	8 bit
MOTOROLA MC 6800	8 bit
MOTOROLA MC 6809	8 bit
ZILOG Z-80	8 bit
NATIONAL LMP-8	8 bit
MOSTECH-6100	8 bit
INTERSIC 6100	12 bit
TOSHIBA TLCS-12	12 bit
GENERAL INSTRUMENT CP1600	16 bit
TITMS	16 bit
DEC-W.D.MCP-1600	16 bit

Characteristics of **second generation** microprocessors:

- 40 pins
- Faster operation
- Larger chip size
- More powerful instructions
- Better interrupt handling capabilities

Applications

- Instrumentation
- Military applications
- Processing control systems
- Complex industrial controls

Third generation microprocessors

The first 16 bit microprocessor capable of addressing of 1 MB memory was introduced by Intel Corporation in the year 1978. Then the same company introduced another 16 bit microprocessor Intel 8088 in the year 1979.

Intel 80816 and 80188 unproved version of Intel 8086 and 8085 respectively.

A few generation microprocessors are listed below:

Microprocessor	Word size
Intel 8086	16bit
Intel 8088	16bit
Intel 80186	16bit
Intel 80188	16bit
Intel 80286	16bit
Zilog 8000	16bit

Motorola 68000	16bit
Motorola 68010	16bit
National ns 16016	16bit

Characteristics of **third generation** microprocessors:

- Provided with 40/48/64 pins
- High speed
- Sizes of internal registers are 8/16/32 bits
- Very short processing capability

Applications

- Advanced common
- Distributed and data processing networks
- Business processing applications

Fourth generation microprocessors

The fourth generation microprocessor was introduced after 1980. It is HCMOS technology. It is 32 bit microprocessor.

Some of the fourth generation microprocessors are given below:

Microprocessor	Word size
Intel 80386	32 bit
Intel 80486	32 bit
Motorola mc 88100	32 bit
Motorola m 68020	32 bit
Motorola m 68030	32 bit
National ns 16032	32 bit

Characteristics:

- Physical memory space of 2^24 byte = 16 MB
- Virtual memory space of 2^40 byte = 1 TB
- Floating point hardware is incorporation

Bit size	Model number	Application
4 bit	HMCS40,COP420 MSC6411,TMS1000, TLCS47 etc	Appliances and toys
8 bit	INTEL8048,INTEL8051 PIC16C56,COP820 PHILIPS7C522,M6805, ROCKWELL6500	Control and monitoring applications

16 bit	H8/532,INTEL80C196 HPC16164	Limited calculations and relatively simple control strategies
32 bit	INTEL80960	To design target robotics, highly intelligent instrumentation , avionics etc

1.5 8051 FLAVORS

The 8051 has the widest range of variants of any embedded controller on the market. The smallest device is the Atmel 89C1051, a 20 Pin FLASH variant with 2 timers, UART, 20mA. The fastest parts are from Dallas, with performance close to 10 MIPS! The most powerful chip is the Siemens 80C517A, with 32 Bit ALU, 2 UARTS, 2K RAM, PLCC84 package, 8 x 16 Bit PWM's, and other features.

Among the major manufacturers are:

AMD	Enhanced 8051 parts (no longer producing 80 × 51 parts)
Atmel	FLASH and semi-custom parts
Dallas	Battery backed, program download, and fastest variants
Intel	8051 through 80c51gb/80c51sl
ISSI	IS80C51/31 runs up to 40MHz
Matra	80c154, low voltage static variants
OKI	80c154, mask parts
Philips	87c748 thru 89c588 - more variants than anyone else
Siemens	80c501 through 80c517a, and SIECO cores
SMC	COM20051 with ARCNET token bus network engine
SSI	80 × 52, 2 × HDLC variant for MODEM use

1.6 SUMMARY

- A microprocessor is a program controlled device(IC) which fetches, decodes and execute instructions.
- Microprocessor is the heart of microcontroller system.
- Microprocessor consists of ALU, register and control circuit.
- A digital computer whose CPU is a microprocessor is called microcomputer.
- The world's first microprocessor Intel 4004 was released by Intel Corporation in the year 1971. The computing system designed using a microprocessor as its CPU is called a micro computer.
- The NMOS process offers faster speed and higher density than PMOS and it is TTL compatible.
- The Intel 8051 is a better hardware architecture single chip microcontroller which was developed by Intel for 1980 for use in embedded systems.
- Intel's original 8051 family was developed using NMOS technology, but later versions, identified by a letter "c" in their names

  (e.g.) 80c51,used CMOS technology.

- The 8057 has the widest range of variants of any embedded controller on the market.
- The 8051 instruction set is optimized for the one bit operations so often desired in real world real-time control application.

1.7 QUESTIONS

1. Microprocessor is the __________ of the microcontroller system.
2. The microprocessor is a __________device.
3. The first generation of microprocessor was introduced in _________ year.
4. What was the basic unit of a microprocessor?
5. A microcontroller normally has which of the following devices on-chip?
   (*a*) RAM (*b*) ROM (*c*) I/O (*d*) All of these

True or false

1. Microcontrollers are normally less expensive than microprocessors.
2. A general purpose microprocessor has on-chip ROM.
3. A microcontroller has on-chip ROM.

Chapter 2

NUMBER SYSTEM

2.1 INTRODUCTION

Everywhere, except for computer related operation the main system of mathematical notation today is the decimal system, which is a base 10 system. As in the other number systems, the position of a symbol in terms of exponential values of the base. That is in the decimal system, the quantity represented by any of the ten symbols used 0, 1, 2, 3, 4, 5, 6, 7, 8, and 9 depends on its position in the number.

Unlike the decimal system, only two digits 0, 1 suffice is representing the number in the binary system. The binary system plays a crucial role in computer science and technology.

2.2 DECIMAL SYSTEM

The base or the radix of a number system is defined as the number of digits it uses to represent the numbers in the system. Since, the decimal number system uses 10 digits 0 to 9 its base or radix is 10, the decimal number system is also called base 10 number system. The weight of each digit of a decimal number system depends on its relative position within the number.

Example:

$$2763 = 2000 + 700 + 60 + 3$$

in other words,

$$2763 = 2 \times 10^3 + 7 \times 10^2 + 6 \times 10^1 + 3 \times 10^0$$

2.3 BINARY SYSTEM

The binary system is a great help in the Nim-like games plainim, Nimble, Turning, Turtle, Scoring, Northcott's game etc. More importantly, the binary system underlies modern technology of electronic digital computers. Computer memory comprises small elements that may only be in two states — OFF / ON that are associated with digits 0 and 1. Such an element is said to represent one bit — binary digit.

For example, 1101

To represent numbers the decimal system uses the powers of 10, whereas the binary system uses in a similar manner the power of 2.

$$(1101)2 = 1.2^3 + 1.2^2 + 0.2 + 1$$

The numbers are different, In fact

$$(1101)_2 = 8 + 4 + 0 + 1 = (13)_{10}$$

Table:

Decimal	Binary
1	1
2	10
3	11
4	100
5	101
6	110
7	111
8	1000
9	1001
10	1010
11	1011
12	1100
13	1101
14	1110
15	1111
16	10000
17	10001
18	10010
19	10011
20	10100

2.4 BINARY ADDITION AND SUBTRACTION

2.4.1 Binary Addition

Binary addition is performed in the same manner as decimal addition. Actually, binary arithmetic is much simpler to learn. The complete table for binary addition as follows,

$$0 + 0 = 0$$

0 + 1 = 1

1 + 0 = 1

1 + 1 = 0 plus a carry over of 1.

Examples of binary addition:

1. 7 + 2

Decimal	Binary
7	111
2	010
9	**1001**

2. 4 + 8

Decimal	Binary
4	100
8	1000
12	**1100**

3. 12 + 15

Decimal	Binary
12	1100
15	1111
27	**11011**

4. 17 + 18

Decimal	Binary
17	10001
18	10010
35	**100011**

2.4.2 Binary Subtraction

Binary subtraction is the inverse operation of addition. The only case in which this occurs with binary numbers is when 1 is subtracted from 0. The remainder is 1, but it is necessary to borrow 1 from the next column to the left. This is the binary subtract table.

0 – 0 = 0

1 – 0 = 1

1 – 1 = 0

0 – 1 = 1 with a borrow of 1.

Examples of binary Subtraction:

1. 7 – 4

Decimal	Binary
7	111
–4	100
3	**011**

2. 18 – 6

Decimal	Binary
18	10010
–6	110
12	**1100**

3. 16 – 4

Decimal	Binary
16	10000
– 4	00100
12	**1100**

4. 12 – 2

Decimal	Binary
12	1100
–2	0010
10	**1010**

2.5 BINARY MULTIPLICATION AND DIVISION

2.5.1 Binary Multiplication

The table for binary multiplication is very short, with only four entries instead of the necessary for decimal multiplication.

$$0 * 0 = 0$$

$$0 * 1 = 0$$

$$1 * 0 = 0$$

$$1 * 1 = 1$$

Examples for binary Multiplication:

1. 4 × 3

Decimal	Binary
4 × 3	100 × 011 100 100 000
12	**01100**

2. 10 × 4

Decimal	Binary
10 × 4	1010 × 100 0000 0000 1010
40	**101000**

3. 102 × 8

Decimal	Binary
102 × 8	1100110 × 1000 0000000 0000000 0000000 1100110
816	**1100110000**

2.5.2 Binary Division

It is a very simple. As in the decimal system (or in any other), division by zero is meaningless. The complete table as,

$$0/1 = 0$$
$$1/0 = 1$$

Examples for binary division:

1. 25 ÷ 5

Decimal	Binary
5 5) 25 25 0	101 101) 11001 101 0101 0101 0

2. 20 ÷ 4

	Decimal		**Binary**
	5		101
4	20	100	10001
	20		101
	0		00100
			00100
			0

3. 18 ÷ 6

	Decimal		**Binary**
	6		110
3	18	011	10010
	18		011
	0		011
			011
			0

4. 29 ÷ 12

	Decimal		**Binary**
	2.416		10.011010101
12	29	1100	11101.00
	24		1100
	50		10100
	48		1100
	20		10000
	12		1100
	80		10000
	72		1100
			……..

2.6 CONVERTING DECIMAL NUMBER TO BINARY

There are several methods for converting a decimal number to a binary number for instance, suppose that you want to convert decimal into corresponding binary number.

For examples of Decimal to Binary:

1.

2	74	
2	37	– 0
2	18	– 1
2	9	– 0
2	4	– 1
2	2	– 0
	1	– 0

$$(74)_{10} = (1001010)_2$$

2.

2	93	
2	46	– 1
2	23	– 0
2	11	– 1
2	5	– 1
2	2	– 1
	1	– 0

$$(93)_{10} = (1011101)_2$$

3.

2	121	
2	60	– 1
2	30	– 0
2	15	– 0
2	7	– 1
2	3	– 1
	1	– 1

$$(121)_{10} = (1111001)_2$$

2.7 CONVERTING BINARY NUMBER TO DECIMAL

The conversion of binary to decimal may be accomplished by using several techniques.

Example of Binary to decimal:

1. $(1 \quad 0 \quad 1)_2$

$1 \times 2^0 = 1 \times 1 = 1$

$0 \times 2^1 = 0 \times 2 = 0$

$1 \times 2^2 = 1 \times 4 = 4$

5

$$(101)_2 = (5)_{10}$$

2. $(1 \quad 0 \quad 1 \quad 1)_2$

$1 \times 2^0 = 1 \times 1 = 1$

$1 \times 2^1 = 1 \times 2 = 2$

$0 \times 2^2 = 0 \times 4 = 0$

$1 \times 2^3 = 1 \times 8 = 8$

11

$$(1011)_2 = (11)_{10}$$

2.8 NEGATIVE NUMBER REPRESENTATION

There are three methods usually employed to represent negative integer in binary system. These methods are known as

(*i*) Signed magnitude method

(*ii*) One's complement method

(*iii*) Two's complement method

2.8.1 The Signed Magnitude Method

In this method the signed magnitude method, the sign bit is assigned the value of 1 to denote that it is a negative number.

Examples of signed magnitude method:

Example 1:

$$(24)_{10} = 11000$$
$$(-24)_{10} = 1\ 11000$$

2	24
2	12 – 0
2	6 – 0
2	3 – 0
	1 – 1

Example 2:

(147) = 10010111

(–147) = 1 10010111

2	147
2	73 – 1
2	36 – 1
2	18 – 0
2	9 – 0
2	4 – 1
2	2 – 0
	1 – 1

2.8.2 One's Complement Method

Subtracting using the 1's complement system is also straight forward. The 1's complement of a number is formed by changing each 1 is the number to a 0 and each 0 is the number to 1.

When subtracting is performed in the 1s complement system, any end-around carry is added to the least significant.

For examples:

1. Binary → 11011   11011
10110 → 1's 01001
00101   1) 00100
→ 1
00101

2. 111011 → 111011
10110 → 1's 101010
100110   1 100101
1
→
0100110

2.8.3 Two's Complement

The 2s complement of a binary number is formed by simply subtracting each digit (bit) of the number from the radix minus one and adding a 1 to the least significant bit. Since, the radix in the binary number system is 2, each bit of the binary number is subtracted from 1. The application of this rule is actually very simple; every 1 in the number is changed to a 0 and every 0 to a 1. Then, a 1 is added to the least significant bit of the number formed.

For examples:

1.

```
        11011   →       11011
     (−)10100   → 1's   01011
                  2's       1
                        -----
                        01100
                        11011
                        -----
Carry is dropped      1 00111
```

2.

```
        11100   →       11100
     (−)00100   →       11011
     --------               1
        11000           -----
                        11100
                        11100
                        -----
Carry is dropped      1|11000
```

2.9 DECIMAL COMPONENTS

In the decimal system the two types are referred to as the 10's complement and 9's complement.

2.9.1 9's Complement

This complement of a decimal number is formed by subtracting each digit of the number from 9.

For examples of 9's complement:

```
(i) 23 – 9's      99
                  23
                  --
                  76

(ii) 48 – 9's     99
                  48
                  --
                  51
```

2.9.2 10's Complement

The 10's complement of any number may be formed by subtracting each digit of the number from 9 and then adding 1 to the least significant digit of the number thus formed.

For example of 10's complement:

Normal subtracts:		**10's complement:**
1. 78		78 – 78
–24		24 – 76
54	carry is dropped	1 | 54

Where,

99
24
———
75
1
———
76

2.10 OCTAL NUMBER SYSTEM

The octal number system has a base or radix; of 8, eight different symbols are used to represent the numbers. There are commonly 0, 1, 2, 3, 4, 5, 6 and 7. An octal number can be converted easily into the binary number with a group of three binary digits replacing each digits of octal system.

The relation between the decimal system, octal system, Hexadecimal and binary system is shown in table.

Decimal	Binary	Octal	HexaDecimal
0	00000	00	0
1	00001	01	1
2	00010	02	2
3	00011	03	3
4	00100	04	4
5	00101	05	5
6	00110	06	6
7	00111	07	7
8	01000	10	8
9	01001	11	9
10	01010	12	A
11	01011	13	B
12	01100	14	C
13	01101	15	D
14	01110	16	E

15	01111	17	F
16	10000	20	10
17	10001	21	11
18	10010	22	12
19	10011	23	13
20	10100	24	14

2.11 CONVERT BINARY TO OCTAL

There is a simple trick for converting a binary number to an octal number simply group the binary digits into groups of 3, starting at the octal point and read each set of three binary digits.

For example of binary to octal:

$$(011\ 100)_2 \rightarrow (34)_8$$
$$(100111000)_2 \rightarrow (470)_8$$
$$(10101)_2 \rightarrow (25)_8$$
$$(1001)_2 \rightarrow (11)_8$$
$$(1011.110)_2 \rightarrow (13.6)_8$$
$$(10101.11)_2 \rightarrow (25.6)_8$$

2.12 CONVERT DECIMAL TO OCTAL

1.

8	487
8	60 – 7
	7 – 4

$(487)_{10} = (747)_8$

2.

8	200
8	25 – 0
	3 – 1

$(200)_{10} = (310)_8$

2.13 HEXADECIMAL

Most mini computers and microcomputers have their memories organized into sets of bytes, each consisting of eight binary digits. Each byte either is used as single entities to represent a single alpha numeric character or is broken into two 4 bit pieces. When the bytes are handles in two 4 bit pieces the programmer is given the option of declaring each 4 bit character as a piece of a binary number or as two, BCD numbers. For instance, the byte 00011000 can be declared a binary number in which case it is equal to 24 decimal, or as two BCD characters in which case it represents the decimal number 18.

When the machine is handling numbers in binary but in groups of four digits it is convenient to have a code for representing each of these sets of four digits. Since, 16 possible different numbers can be represented, the digits 0 through 9 will not suffice, so the letters A, B, C, D, E and F are used.

2.14 CONVERT BINARY TO HEXADECIMAL

Example:

$$1010/1110/1101 \rightarrow (AED)_{16}$$

$$010/1010_2 = 2A_{16}$$

$$1010/0010 = A2_{16}$$

2.15 CONVERT HEXADECIMAL TO DECIMAL

1. A B 6

$$6 \times 16^0 = 6 \times 1 = 6$$

$$11 \times 16^1 = 11 \times 16 = 176$$

$$10 \times 16^2 = 10 \times 256 = 2560$$

2742

$$\mathbf{AB6_{16} = (2742)_{10}}$$

2. **(1 A 2 E)2**

$$14 \times 16^0 = 14 \times 1 = 14$$

$$2 \times 16^1 = 2 \times 16 = 32$$

$$A \times 16^2 = 10 \times 256 = 2560$$

$$1 \times 16^3 = 1 \times 4096 = 4096$$

6702

$$\mathbf{(1A2E)_{16} = (6702)_{10}}$$

2.16 EXCESS – 3 CODE

The excess 3 code is another important BCD code. This code has been derived from the standard BCD code. The excess -3 code for the decimal numbers is obtained by adding 0011 to the code in standard BCD is 0110. We get the excess 3 code for 6 by adding 0011 to 0110 that is the excess 3 code for 6 is 1001. The excess 3 code is given in table

Decimal	Excess-3 Code
0	0001
1	0100
2	0101
3	0110
4	0111
5	1000
6	1001
7	1010
8	1011
9	1100
10	0100 0011
20	0101 0011
99	1100 1100
100	0100 0011 0011

2.17 GRAY CODE

This code is a unit distance code and a non-weighted code. It is not suitable for arithmetic operations but it is very useful for transducers line shaft position encoders, Input output devices, analog to digital converters and other peripheral equipments.

The relation between the gray code and the natural binary is shown in the table.

Decimal	Binary	Gray Code
1	0000	0000
2	0001	0001
3	0010	0011
4	0100	0110
5	0101	0111
6	0110	0101

7	0111	0100
8	1000	1100
9	1001	1101
10	1010	1111
11	1011	1110
12	1100	1010
13	1101	1011
14	1110	1001
15	1111	1000

2.18 SUMMARY

- The individual memory cells used in computers are in bistable in operation and capable of storing a single binary bit.
- It is most practical to use the binary number system to represent numbers and the system was explained along with conversion techniques to and from decimal.
- Negative numbers are represented in computer by using a sign bit which is a 1 when the number is negative and a 0 for positive numbers.
- Negative numbers are often represented by using 1's are 2's complement form.
- The direct representation of decimal numbers can be accomplished by using a binary coded decimal representation.
- The octal and hexadecimal number systems were described.
- These are useful in representing binary numbers in a compact form and to facilitate communication of values in written presentation.
- Computers are often organized with numbers represented in groups of 8 bit which makes hexadecimal particularly useful at the time.

2.19 QUESTIONS

Fill up the blanks

1. Binary 1010 in decimal system is equivalent to ______________.
2. The decimal equivalent of the binary number 10110.0101011101 is__________.
3. Binary 101010 is equivalent to decimal number is __________.
4. The hexadecimal number A492 is equivalent to decimal number is_____________.
5. The binary number 011 1011011 is equivalent to decimal number ____________
6. The digit 0 with carry of 1 is the sum of binary addition is __________.
7. In BCD code is represented as __________
8. Octal 16 is equal to decimal _________.

9. The sum of $(111010)_2$ and $(11011)_2$ in decimal form will be__________
10. $(100101)_2$ is ____________

True or false:

1. As compared to digital computers, micro computers have high cost and big size.
2. The number of binary digits that make up the word is the word length.
3. A byte is an 8 binary digit word length.
4. BCD numbers express each decimal digit as a byte.
5. For the binary number 101101110 the equivalent octal number is 556.

Chapter 3

The 8051 Architecture

3.1 INTRODUCTION

The 8051 is an 8 bit microcontroller originally developed by Intel in 1980. It is the world's popular microcontroller core made by many independent manufacturer truly multi sourced. There were 126 million 8051s and variants shipped in 1993. A typical 8051 contains.

- CPU with Boolean processor
- 5 or 6 interrupt. 2 are external and 2 are priority levels.
- Programmable full duplex serial port baud rate provided by one of the timers.
- 32 input/output lines (four 8 bit ports)
- RAM
- ROM/EPROM in some models

One strong point of the 8051 is the way it handles interrupts vectoring to fixed 8 byte areas is convenient and efficient. Most interrupts routines are very short and generally can fit in to the 8 byte area. The 8051 instructions set is optimized for the one bit operations so often desired in real world, real time control applications. Bit addressing can be used for test pin monitoring or program control flags.

3.2 MICROCONTROLLER STANDARD

Microcontrollers' producers have been struggling for a long time for attracting more and more choosy customers. Every couple of days a new chip with a higher operating frequency, more memory and more high-quality A/D converters comes on the market.

Nevertheless, by analyzing their structures it is concluded that most of them have the same (or at least very similar) architecture known in the product catalogs as "8051 compatible".

The whole story began in the far 80s when Intel launched its series of the microcontrollers labeled with MCS 051. Although, several circuits belonging to this series had quite modest features in comparison to the new ones, they took over the world very fast and became a standard for what nowadays is meant by a word microcontroller.

The reason for success and such a big popularity is a skillfully chosen configuration which satisfies needs of a great number of the users allowing at the same time stable expanding (refers to the new types of the microcontrollers). Besides, since a great deal of software has been developed in the meantime, it simply was not profitable to change anything in the microcontroller's basic core. That is the reason for having a great number of various microcontrollers which actually are solely upgraded versions of the 8051 family.

8051 Microcontroller has Nothing Impressive at First Sight:

- 4 kb program memory is not much at all
- 128 kb ram satisfies basic needs, but it is not imposing amount
- 4 ports having in total of 32 input/output lines are mostly enough to make connections to peripheral environment and are not luxury to all

The 8051 Microcontroller have Nothing Impressive at First Sight:

The whole configuration is obviously envisaged as such to satisfy the needs of most programmers who work on development of automation device. One of advantage of microcontroller is that nothing in missing nothing is too much. In other words it is created exactly in advanced to the average users states and needs. The other advantage is the way ram is organized, the way central processor unit operates and ports which maximally use all recourses and enable further upgrading.

3.3 THE FEATURES OF THE 8051

The features of the 8051 family as follows:

- 8 bit CPU optimized for control application
- 4096 bytes on chip program memory
- 128 bytes on chip data memory
- 64 k program memory address space
- 64 k data memory address space
- 32 bidirectional and individual addressable i/o lines
- Two 16 bit timer/counter
- Full duplex UART
- Extensive Boolean processing capabilities
- Two level prioritized interrupt structure
- Direct byte and bit address ability

- Binary on decimal arithmetic
- On chip clock oscillator
- One microseconds instructions cycle with 12MHz crystal
- Hardware multiple and divide in 4 microseconds.

3.4 8051 MICROCONTROLLER HARDWARE

3.4.1 8051 Architecture

The figure shows the internal block diagram of 8051. It consists of a CPU two kinds of memory sections, input/output ports, specify function register and control logic needed for a variety of peripheral functions. These elements communicate though an eight bit data bus which runs throughout the chip referred as internal data bus. This bus is buffered to the outside world through an I/O port when memory or I/O expansion is desired.

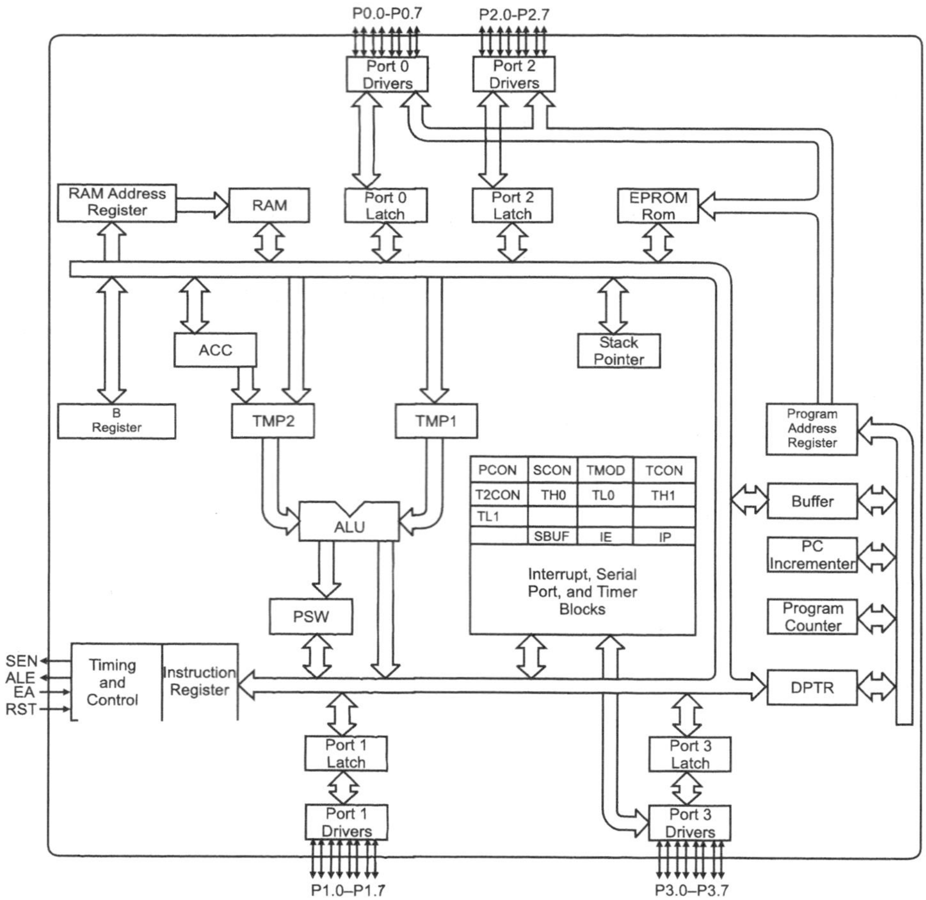

Fig. 3.1 8051 Architecture

3.4.2 Pin-out Diagram of 8051

The 8051 is packaged in a 40 pin dip. The figure shows the pin diagram of 8051. It is important to note that many pins of 8051 are used for more than one function. The alternative functions of pins are shown is bold letters.

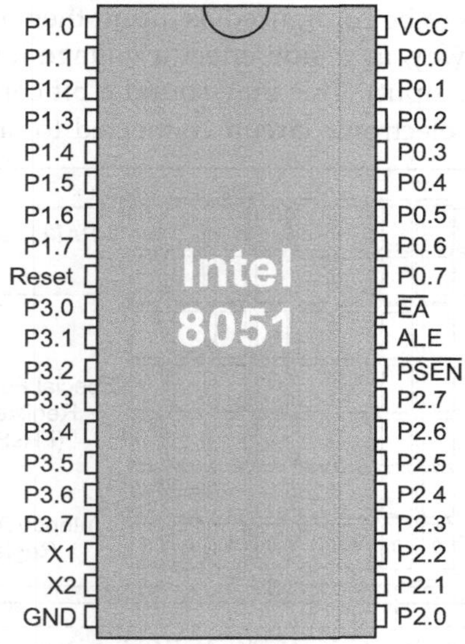

Fig. 3.2 Pin-out diagram of 8051

3.4.3 8051 Oscillator and Clock

The 8051 requires the existence of an external oscillator circuit. The oscillator circuit usually around 12 MHz although the 8051 is capable of running at a maximum 40 MHz. Each machine cycle in the 8051 is 12 clock cycles, giving an effective cycle rate at 1 MHz to 3.33MHz.

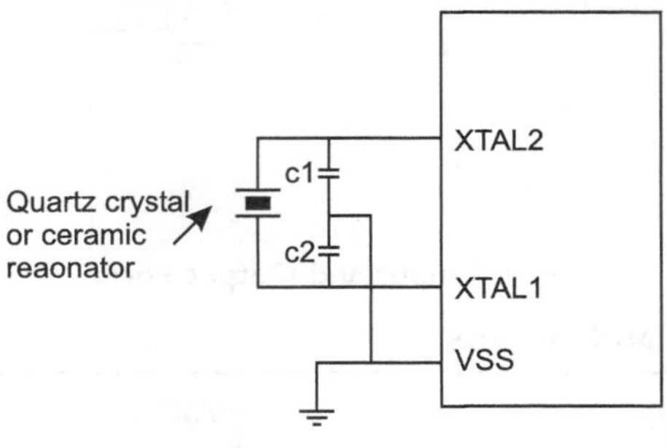

Fig. 3.3 Using the on chip oscillator

The circuit that generates cock pulses, to synchronizes all the internal operations is the heart of the 8051.XTALI and XTAL pins are used to connecting a resonant network to form a oscillator basic internal clock frequency is the crystal from 1 MHz to 16 MHz the figure shows the detail of the oscillator and clock circuit.

3.4.4 Input/Output Ports (I/O Ports)

All 8051 microcontrollers have 4 I/O ports, each consisting of 8 bits which can be configured as inputs or outputs. This means that the user has on disposal in total of 32 input/output lines connecting the microcontroller to peripheral devices.

A logic state on a pin determines whether it is configured as input or output: 0 = output, 1 = input. If a pin on the microcontroller needs to be configured as output, then logic zero (0) should be applied to the appropriate bit on I/O port. In this way, a voltage level on the appropriate pin will be 0.

Similar to that, if a pin needs to be configured as input, then a logic one (1) should be applied to the appropriate port. In this way, as a side effect a voltage level on the appropriate pin will be 5V (as it is case with any TTL input). This may sound a bit confusing but everything becomes clear after studying a simplified electronic circuit connected to one I/O pin.

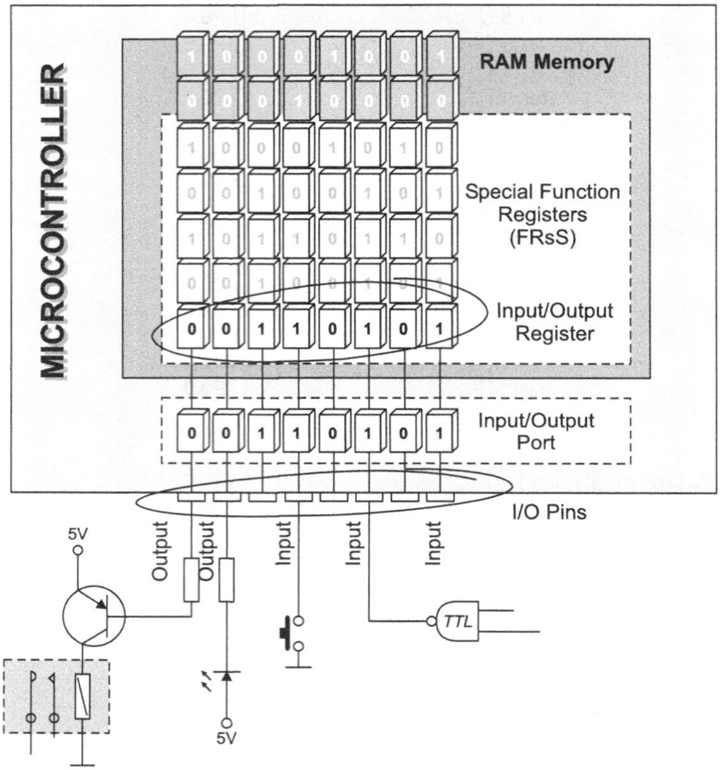

Fig. 3.4 Input and Output Ports

Input/Output (I/O)pin, ports and circuits

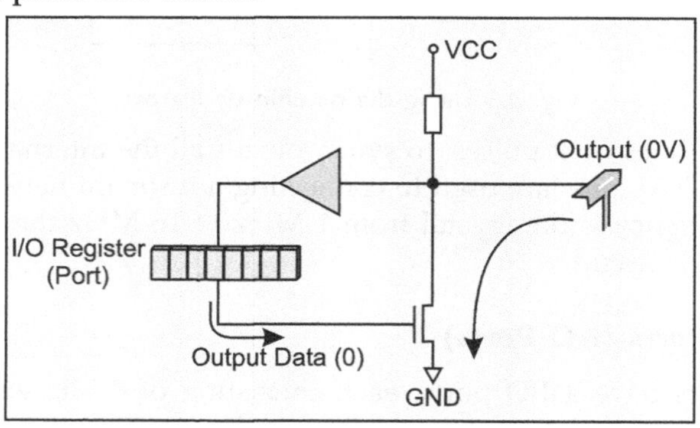

Fig. 3.5 Input/Output Ports and Circuits

This is a simplified overview of what is connected to a pin inside the microcontroller. It concerns all pins except those included in P0 which do not have embedded pull-up resistor.

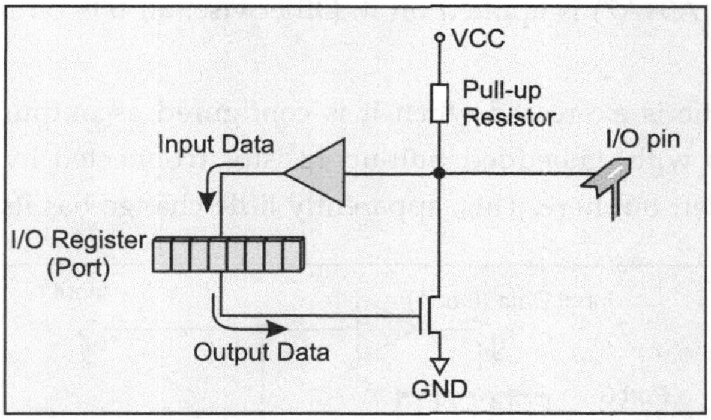

Fig. 3.6 Input/Output Register

Output pin

Logic zero (0) is applied to a bit in the P register. By turning output FE transistor on, the appropriate pin is directly connected to ground.

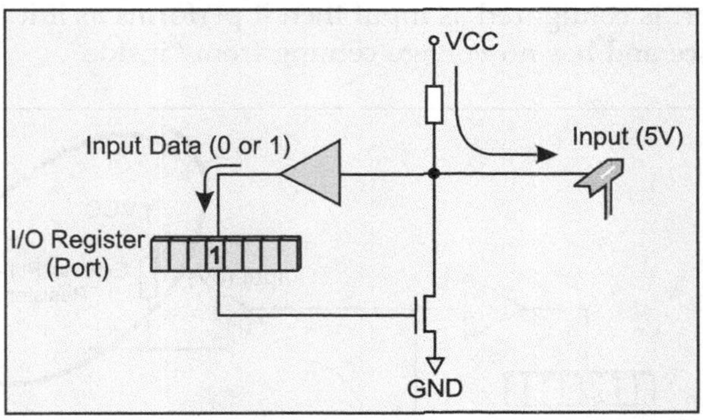

Fig. 3.7 Port 1

Input pin

Logic one (1) is applied to a bit in the P register. Output FE transistor is turned off. The appropriate pin remains connected to voltage power supply through a pull-up resistor of high resistance.

A logic state (voltage) on any pin can be changed or read at any moment. A logic zero (0) and logic one (1) are not equal. Logic one (0) represents almost short circuit to ground. Such a pin is configured as output.

A logic one (1) is "loosely" connected to voltage power supply through resistors of high resistance. Since, this voltage can be easily "pulled down" by an external signal, such a pin is configured as input.

Port 0

It is specific to this port to have a double purpose. If external memory is used then the lower address byte (addresses A0–A7) is applied on it. Otherwise, all bits on this port are configured as inputs or outputs.

Another characteristic is expressed when it is configured as output. Namely, unlike other ports consisting of pins with embedded pull-up resistor (connected by its end to 5 V power supply), this resistor is left out here. This, apparently little change has its consequences:

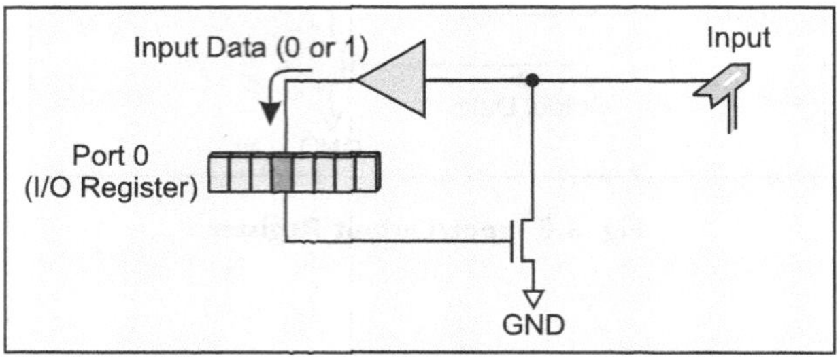

Fig. 3.8 Port 0

If any pin on this port is configured as input then it performs as if it "floats". Such input has unlimited input resistance and has no voltage coming from "inside".

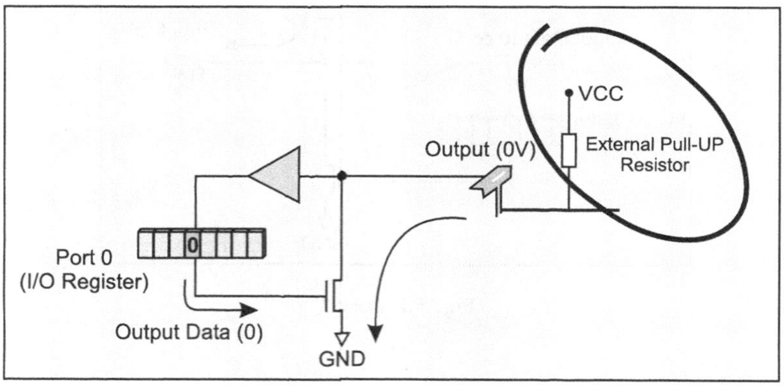

Fig. 3.9 Port 0 (Data Flow)

When the pin is configured as output, it performs as "open drain", meaning that by writing 0 to some port's bit, the appropriate pin will be connected to ground (0V). By writing 1, the external output will keep on "floating". In order to apply 1 (5V) on this output, an external pull-up resistor must be embedded.

Note:

Only in case P0 is used for addressing external memory (only in that case), the microcontroller will provide internal power supply source in order to establish logical ones on pins. There is no need to add external pull-up resistors.

Port 1

This is a true I/O port, because there are no role assigning as it is the case with P0. Since, it has embedded pull-up resistors it is completely compatible with TTL circuits.

Port 2

Similar to P0, when using external memory, lines on this port occupy addresses intended for external memory chip. This time it is the higher address byte with addresses A8–A15. When there is no additional memory, this port can be used as universal input-output port similar by its features to the port 1.

Port 3

Even though all pins on this port can be used as universal I/O port, they also have an alternative function. Since each of these functions use inputs, then the appropriate pins have to be configured like that. In other words, prior to using some of reserve port functions, a logical one (1) must be written to the appropriate bit in the P3 register. From hardware's perspective, this port is also similar to P0, with the difference that its outputs have a pull-up resistor embedded.

Current limitations on pins

When configured as outputs (logic zero (0)), single port pins can "receive" current of 10mA. If all 8 bits on a port are active, total current must be limited to 15mA (port P0: 26mA). If all ports (32 bits) are active, total maximal current must be limited to 71mA. When configured as inputs (logic 1), embedded pull-up resistor provides very weak current, but strong enough to activate up to 4 TTL inputs from LS series.

It may be seen from description of some ports, that even though all pins have more or less similar internal structure, it is necessary to pay attention to which of them will be used for what and how.

For example: If they are used as outputs with high voltage level (5V), then port 0 should be avoided because its pins do not have added resistor for connection to +5V. Only low logic level can be obtained therefore, if another port is used for the same purpose, one should have in mind that pull-up resistors have a relatively high resistance. Consequently, it can be counted on only several hundreds microamperes of current coming out of a pin.

3.4.5 DPTR Register (Data Pointer)

These registers are not true ones, because they do not physically exist. They consist of two separate registers DPH (data pointer high) and (data pointer low). Their 16 bits are used for external memory addressing. They may be handled as a 16 bit register or as two independent 8 bit registers. The DPTR registers are usually used for storing data and immediate results which have nothing to do with memory locations.

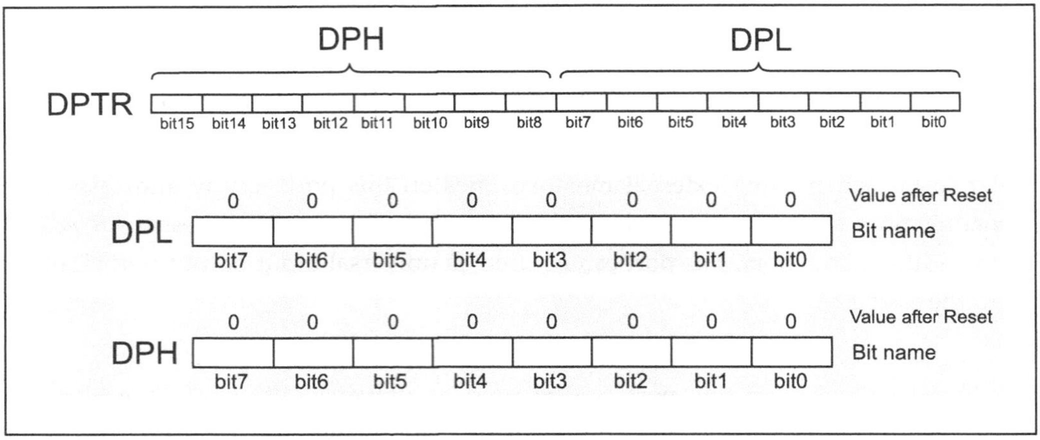

Fig. 3.10 Data Pointer

3.4.6 Program Counter

The 8051 has a 16-bit program counter. It is used to hold the address of memory location from which the next instruction is to be fetched. Due to this the width of the program counter decides the maximum program length in bytes. For example, 8051 is 16-bit wide. This means that the 8051 can access program address 0000 to FFFFH, a total of 64K bytes of code.

The PC is automatically incremented to point the next instruction in the program sequence after execution of the current instruction. It may also be altered by certain instructions. The PC is the only register that does not have an internal address.

3.4.7 Register Set

Register A accumulator:

It is an 8 bit register. It holds a source operand and receives the result of the arithmetic instructions (Addition, Subtraction, Multiplication and Division). The accumulator can be the source or destination for logical operations and a number of special data movement instructions. Includes lookup tables and external RAM expansion. Several functions apply exclusively to the accumulator rotate parity computation, testing for zero and so on.

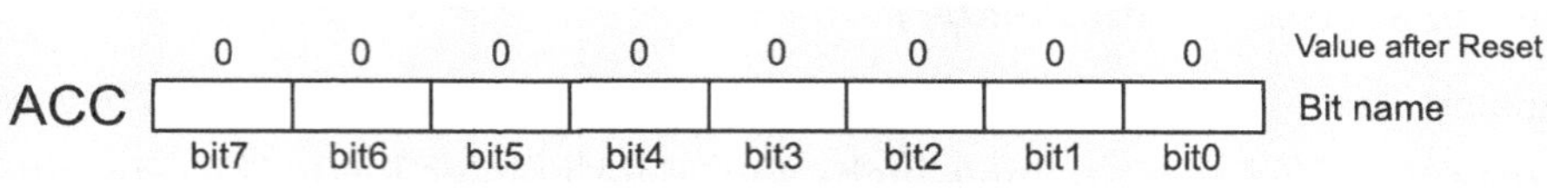

Fig. 3.11 Accumulator

B registers:

B register is used during multiply and divide operations which can be performed only upon numbers stored in the A and B registers. All other instructions in the program can use this register as a spare accumulator (A).

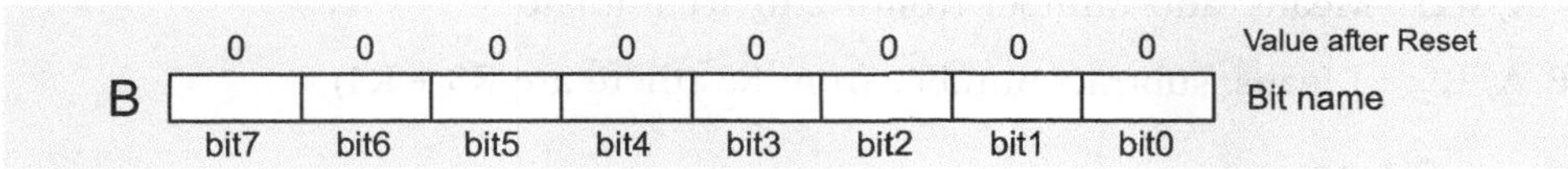

Fig. 3.12 B Register

Note: During programming, each of registers is called by name so that their exact addresses are not so important for the user. During compiling into machine code (series of hexadecimal numbers recognized as instructions by the microcontroller), PC will automatically, instead of registers' name, write necessary addresses into the microcontroller.

R Registers (R0–R7)

This is a common name for the total 8 general purpose registers (R0, R1, R2 ... R7). Even they are not true SFRs, they deserve to be discussed here, because of their purposes. The bank is active when the R registers are in use. Similar to the accumulator, they are used for temporary storing variables and intermediate results. Which of the banks will be active it depends on two bits included in the PSW Register. These registers are stored in four banks in the scope of RAM.

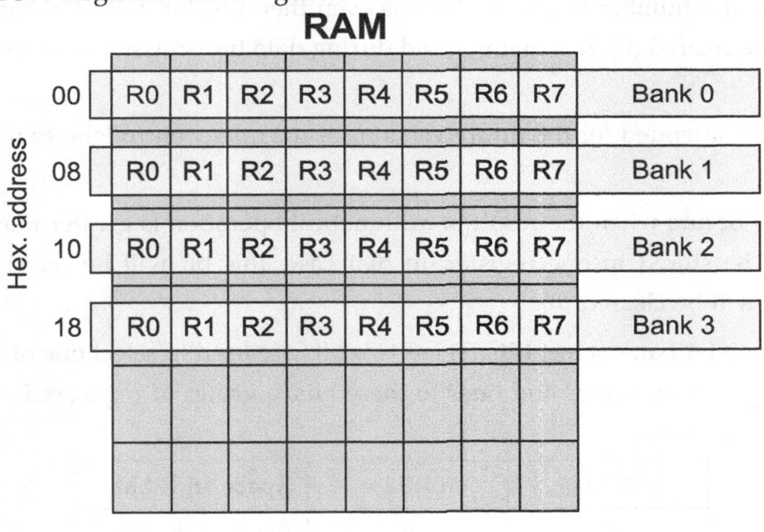

Fig. 3.13 RAM

The following example best illustrates the useful purpose of these registers. Suppose that mathematical operations on numbers previously stored in the R registers should be performed: (R1 + R2) – (R3 + R4). Obviously, a register for temporary storing results of addition is needed. Everything is quite simple and the program is as follows:

MOV A, R3; Means: move number from R3 into accumulator

ADD A, R4; Means: add number from R4 to accumulator (result remains in accumulator)

MOV R5, A; Means: temporarily move the result from accumulator into R5

MOV A, R1; Means: move number from R1 into accumulator

ADD A, R2; Means: add number from R2 to accumulator

SUBB A, R5; Means: subtract number from R5 (there are R3 + R4)

3.4.8 Flags and PSW

PSW Register (Program Status Word)

PSW	bit7	bit6	bit5	bit4	bit3	bit2	bit1	bit0
Value after Reset	0	0	0	0	0	0	0	0
Bit name	CY	AC	F0	RS1	RS0	OV		P

Fig. 3.14 Program Status Word

This is one of the most important SFRs. The Program Status Word (PSW) contains several status bits that reflect the current state of the CPU. This register contains: Carry bit, Auxiliary Carry, two register bank select bits, Overflow flag, parity bit, and user-definable status flag. The ALU automatically changes some of register's bits, which is usually used in regulation of the program performing.

P – Parity bit. If a number in accumulator is even then this bit will be automatically set (1), otherwise it will be cleared (0). It is mainly used during data transmission and receiving via serial communication.

Bit 1. This bit is intended for the future versions of the microcontrollers, so it is not supposed to be here.

OV Overflow occurs when the result of arithmetical operation is greater than 255 (decimal), so that it cannot be stored in one register. In that case, this bit will be set (1). If there is no overflow, this bit will be cleared (0).

RS0, RS1 - Register bank select bits. These two bits are used to select one of the four register banks in RAM. By writing zeroes and ones to these bits, a group of registers R0–R7 is stored in one of four banks in RAM.

RS1	RS2	Space in RAM
0	0	Bank0 00h–07h
0	1	Bank1 08h–0Fh
1	0	Bank2 10h–17h
1	1	Bank3 18h–1Fh

F0 - Flag 0. This is a general-purpose bit available to the user.

AC - Auxiliary Carry Flag is used for BCD operations only.

CY - Carry Flag is the (ninth) auxiliary bit used for all arithmetical operations and shift instructions.

3.4.9 Stack and Stack Pointer

The stack refers to an area of internal RAM that is used in conjunctions with certain opcodes data stored and retrieve data quickly. The stack pointer register is used by the 8051 to hold an internal RAM address that is called top of stack.

The stack pointer register is 8 bit wide. It is incremented before data is stored during PUSH and CALL instructions and decremented after data is restored during POP and RET instructions. The stack array can reside anywhere in on chip RAM. The stack pointer is initialized to 07H after a reset. This causes the stack to begin at locations 08H. The operation of stack and stack pointer is illustrated in figure.

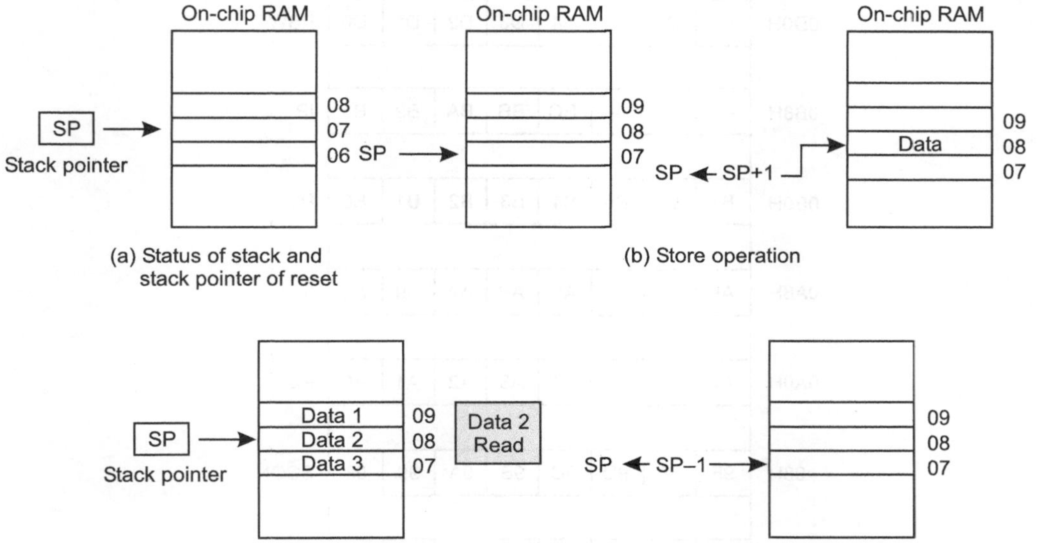

Fig. 3.15 Stack and Stack Pointer

3.4.10 Special Function Registers

SFRs are a kind of control table used for running and monitoring microcontroller's operating. Each of these registers, even each bit they include, has its name, address in the scope of RAM and clearly defined purpose (for example: timer control, interrupt, serial connection etc.). Even though there are 128 free memory locations intended for their storage, the basic core, shared by all types of 8051 controllers, has only 21 such registers. Rest of locations is intentionally left free in order to enable the producers to further improved models keeping at the same time compatibility with the previous versions. It also enables the use of programs written a long time ago for the microcontrollers which are out of production now.

8051 uses memory mapped i/o through a set of special function register that are implemented in the address space immediately above the 128 bytes of RAM. Figure shows special function bit address. All access to the four I/O ports, the CPU registers, interrupt control registers, the timer/ counter, UART and power control are performed through registers between 80H and FFH.

Direct Byte Address	Bit Address (MSB)							(LSB)	Hardware Register Symbol
0FFH									
0F0H	F7	F6	F5	F4	F3	F2	F1	F0	B
0E0H	E7	E6	E5	E4	E3	E2	E1	E0	ACC
0D0H	D7	D6	D5	D4	D3	D2	D1	D0	PSW
0B8H	---	---	---	BC	BB	BA	B9	B8	IP
0B0H	B7	B6	B5	B4	B3	B2	B1	B0	P3
0A8H	AF	---	---	AC	AB	AA	A9	A8	IE
0A0H	A7	A6	A5	A4	A3	A2	A1	A0	P2
98H	9F	9E	9D	9C	9B	9A	99	98	SCON
90H	97	96	95	94	93	92	91	90	P1
88H	8F	8E	8D	8C	8B	8A	89	88	TCON
80H	87	86	85	84	83	82	81	80	P0

Fig. 3.16 Special Function Register

3.5 8051 MICROCONTROLLER MEMORY ORGANIZATION

The microcontroller memory is divided into Program Memory and Data Memory. Program Memory (ROM) is used for permanent saving program being executed, while Data Memory (RAM) is used for temporarily storing and keeping intermediate results and variables. Depending on the model in use (still referring to the whole 8051 microcontroller family) at most a few Kb of ROM and 128 or 256 bytes of RAM can be used. However...

All 8051 microcontrollers have 16-bit addressing bus and can address 64 kb memory. It is neither a mistake nor a big ambition of engineers who were working on basic core development. It is a matter of very clever memory organization which makes these controllers a real "programmers' tidbit".

Program Memory

The oldest models of the 8051 microcontroller family did not have internal program memory. It was added from outside as a separate chip. These models are recognizable by their labels beginning with 803 (for ex. 8031 or 8032). All later models have a few Kbytes ROM embedded, Even though it is enough for writing most of the programs, there are situations when additional memory is necessary. A typical example of it is the use of so called lookup tables. They are used in cases when something is too complicated or when there is no time for solving equations describing some process. The example of it can be totally exotic (an estimate of self-guided rockets' meeting point) or totally common (measuring of temperature using non-linear thermo element or asynchronous motor speed control). In those cases, all needed estimates and approximates are executed in advance and the final results are put in the tables (similar to logarithmic tables).

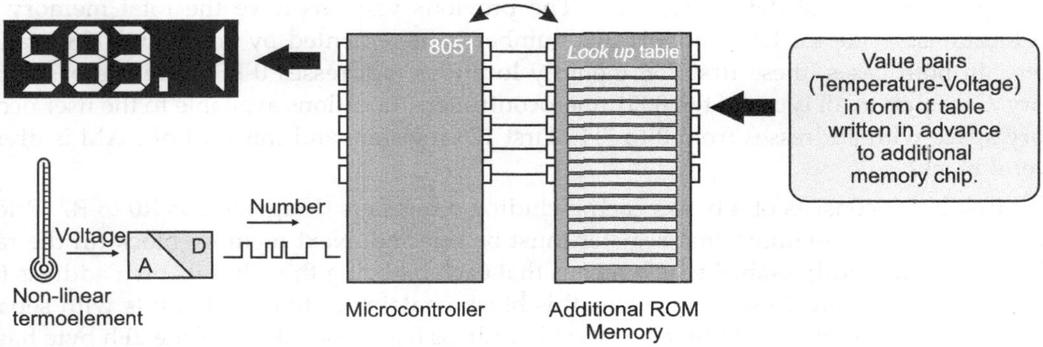

Fig. 3.17 Program Memory

How does the microcontroller handle external memory depend on the pin EA logic state:

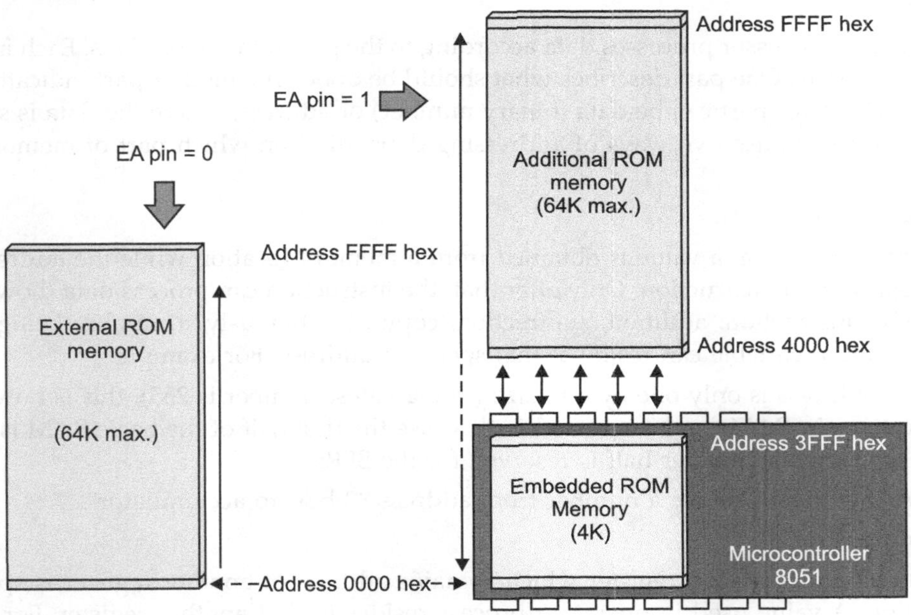

Fig. 3.18 Additional Memory

EA = 0 In this case, internal program memory is completely ignored, only a program stored in external memory is to be executed.

EA = 1 In this case, a program from builtin ROM is to be executed first (to the last location). Afterwards, the execution is continued by reading additional memory.

In both cases, P0 and P2 are not available to the user, because they are used for data and address transmission. Besides, the pins ALE and PSEN are used too.

Data Memory

As already mentioned, Data Memory is used for temporarily storing and keeping data and intermediate results created and used during microcontroller's operating. Besides, this microcontroller family includes many other registers such as: hardware counters and timers, input/output ports, serial data buffers etc. The previous versions have the total memory size of 256 locations, while for later models this number is incremented by additional 128 available registers. In both cases, these first 256 memory locations (addresses 0-FFh) are the base of the memory. Common to all types of the 8051 microcontrollers. Locations available to the user occupy memory space with addresses from 0 to 7Fh. First 128 registers and this part of RAM is divided in several blocks.

The first block consists of 4 banks each including 8 registers designated as R0 to R7. Prior to access them, a bank containing that register must be selected. Next memory block (in the range of 20h to 2Fh) is bit- addressable, which means that each bit being there has its own address from 0 to 7Fh. Since, there are 16 such registers, this block contains in total of 128 bits with separate addresses (The 0th bit of the 20h byte has the bit address 0 and the 7th bit of the 2Fh byte has the bit address 7Fh). The third group of registers occupies addresses 2Fh-7Fh (in total of 80 locations) and does not have any special purpose or feature.

Addressing

While operating, processor processes data according to the program instructions. Each instruction consists of two parts. One part describes what should be done and another part indicates what to use to do it. This later part can be data (binary number) or address, where the data is stored. All 8051 microcontrollers use two ways of addressing depending on which part of memory should be accessed:

Direct Addressing

On direct addressing, a value is obtained from a memory location while the address of that location is specified in instruction. Only after that, the instruction can process data (how depends on the type of instruction: addition, subtraction, copy...). Obviously, a number being changed during operating a variable can reside at that specified address. For example:

Since, the address is only one byte in size (the greatest number is 255), this is how only the first 255 locations in RAM can be accessed in this case the first half of the basic RAM is intended to be used freely, while another half is reserved for the SFRs.

MOV A,33h; Means: move a number from address 33 hex. to accumulator

Indirect Addressing

On indirect addressing, a register which contains address of another register is specified in the instruction. A value used in operating process resides in that another register. For example: Only RAM locations available for use are accessed by indirect addressing (never in the SFRs). For all latest versions of the microcontrollers with additional memory block (those 128 locations in Data Memory), this is the only way of accessing them. Simply, when during operating, the instruction including "@" sign is encountered and if the specified address is higher than 128 (7F hex.), the processor knows that indirect addressing is used and jumps over memory space reserved for the SFRs.

MOV A, @R0; Means: Store the value from the register whose address is in the R0 register into accumulator

On indirect addressing, the registers R0, R1 or Stack Pointer are used for specifying 8-bit addresses. Since, only 8 bits are avilable, it is possible to access only registers of internal RAM in this way (128 locations in former or 256 locations in latest versions of the microcontrollers). If memory extension in form of additional memory chip is used then the 16-bit DPTR Register (consisting of the registers DPTRL and DPTRH) is used for specifying addresses. In this way it is possible to access any location in the range of 64K.

3.6 INTERNAL MEMORY

The 8051 has internal RAM and ROM memory. Additional memory can be added externally using suitable circuits.

3.6.1 Internal RAM

First 128 register this part of ram is divided in several blocks.

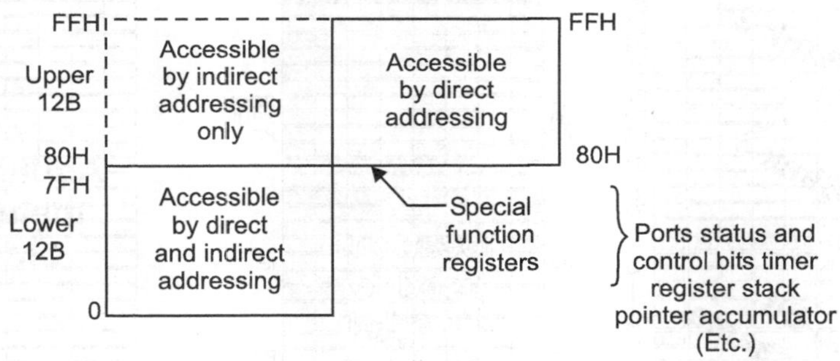

Fig. 3.19 Internal RAM

The first bank consists of four banks each including 8 registers designated as R0–R7, the four register banks are bank0, bank1, bank2 and bank3.location available to the user occupy memory space with addresses from 00–FF.

Next memory block is bit addressable which means that each bit being there has its own address from 00–7FH. Since, there are 16 such registers, this block contain in total of 128 bits with separate addresses.

The third group of registers occupies addresses 2fh–7fh and does not have any special purpose or features.

3.6.2 Additional Memory Block of Data Memory

In order to satisfy the programmers' permanent hunger for Data Memory, producers have embedded an additional memory block of 128 locations into the latest versions of the 8051 microcontrollers. Naturally, it's not so simple. The problem is that electronics performing addressing has 1 byte (8 bits) on disposal and due to that it can reach only the first 256 locations.

In order to keep already existing 8-bit architecture and compatibility with other existing models a little trick has been used.

Using trick in this case means that additional memory block shares the same addresses with existing locations intended for the SFRs (80h–FFh). In order to differentiate between these two physically separated memory spaces, different ways of addressing are used. A direct addressing is used for all locations in the SFRs, while the locations from additional RAM are accessible using indirect addressing.

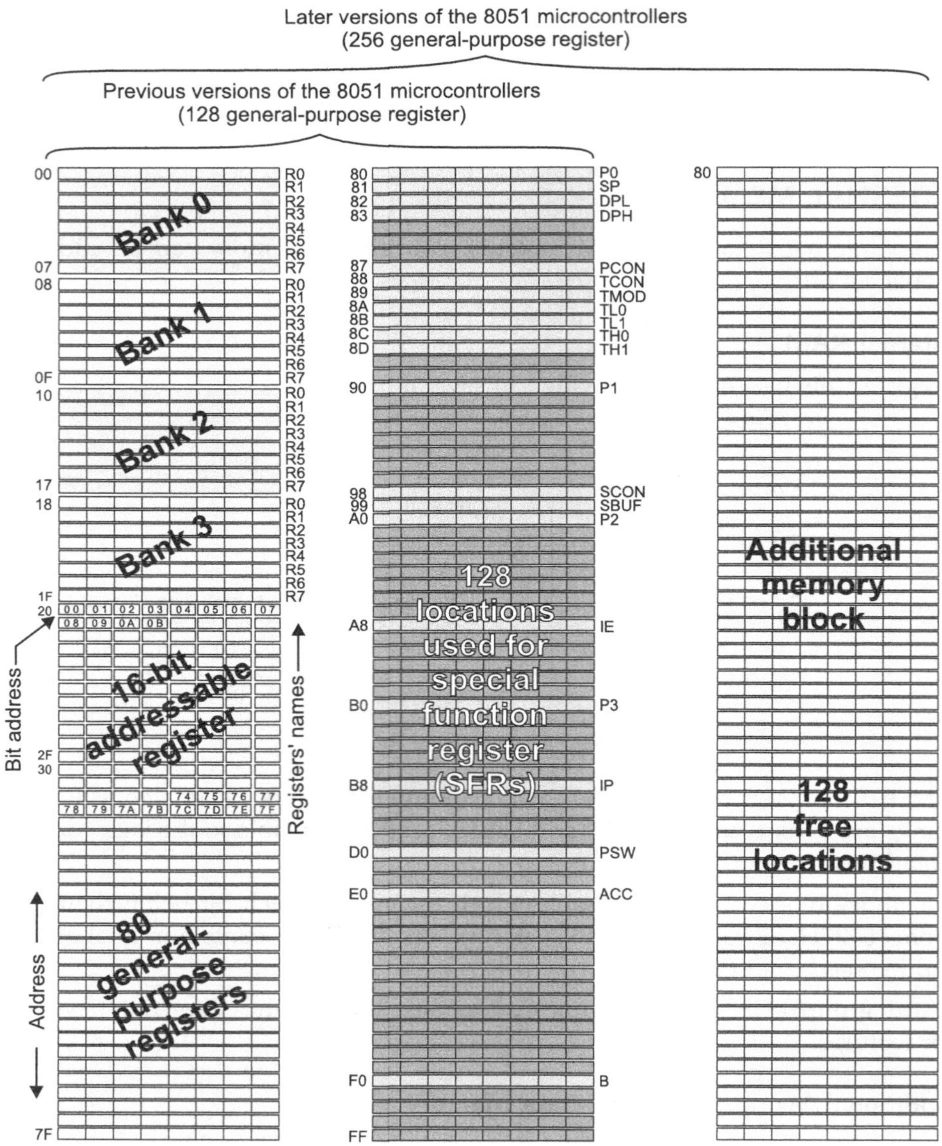

Fig. 3.20 Additional Memory Block

In case on-chip memory is not enough, it is possible to add two external memory chips with capacity of 64Kb each. I/O ports P2 and P3 are used for their addressing and data transmission.

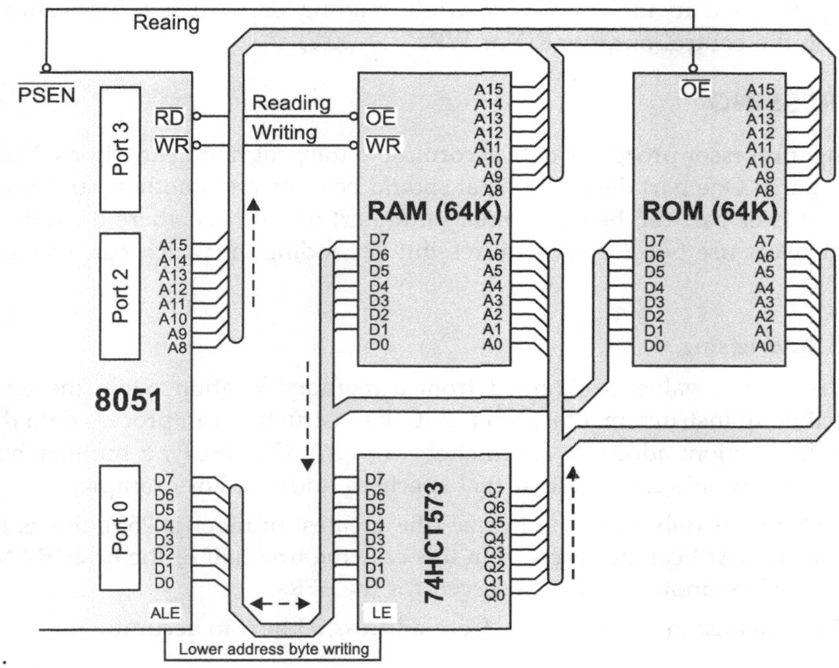

Fig. 3.21 RAM and ROM Memory

From the users' perspective, everything functions quite simple if properly connected, because the most operations are performed by the microcontroller itself. The 8051 microcontroller has two separate reading signals RD# (P3.7) and PSEN#. The first one is activated byte from external data memory (RAM) should be read, while another one is activated to read byte from external program memory (ROM). These both signals are active at logical zero (0) level. A typical example of such memory extension using special chips for RAM and ROM is shown on the previous picture. It is called *Hardware architecture*.

Even though the additional memory is rarely used with the latest versions of the microcontrollers, it will be described here in short what happens when memory chips are connected according to the previous schematic. It is important to know that the whole process is performed automatically, *i.e.*, with no intervention in the program.

- When the program during execution encounters the instruction which resides in external memory (ROM), the microcontroller will activate its control output ALE and set the first 8 bits of address (A0–A7) on P0. In this way, IC circuit 74HCT573 which "lets in" the first 8 bits to memory address pins is activated.
- A signal on the pin ALE closes the IC circuit 74HCT573 and immediately afterwards 8 higher bits of address (A8–A15) appear on the port. In this way, a desired location in additional program memory is completely addressed. The only thing left over is to read its content.
- Pins on P0 are configured as inputs, the pin PSEN is activated and the microcontroller reads content from memory chip. The same connections are used both for data and lower address byte.

Similar occurs when it is a needed to read some locations from external Data Memory. Now, addressing is performed in the same way, while reading or writing is performed via signals which appear on the control outputs RD or WR.

3.7 ADDRESSING

While operating, processor processes data according to the program instructions. Each instruction consists of two parts. One part describes what should be done and another part indicates what to use to do it. This later part can be data (binary number) or address where the data is stored. All 8051 microcontrollers use two ways of addressing depending on which part of memory should be accessed.

3.7.1 Direct Addressing

On direct addressing, a value is obtained from a memory location while the address of that location is specified in instruction. Only after that, the instruction can process data (how depends on the type of instruction: addition, subtraction, copy...). Obviously, a number being changed during operating a variable can reside at that specified address. For example:

Since, the address is only one byte in size (the greatest number is 255), this is how only the first 255 locations in RAM can be accessed in this case the first half of the basic RAM is intended to be used freely, while another half is reserved for the SFRs.

MOV A, 33h; Means: move a number from address 33 hex. to accumulator

3.7.2 Indirect Addressing

On indirect addressing, a register which contains address of another register is specified in the instruction. A value used in operating process resides in that another register.

For example:

Only RAM locations available for use are accessed by indirect addressing (never in the SFRs). For all latest versions of the microcontrollers with additional memory block (those 128 locations in Data Memory), this is the only way of accessing them. Simply, when during operating, the instruction including "@" sign is encountered and if the specified address is higher than 128 (7F hex.), the processor knows that indirect addressing is used and jumps over memory space reserved for the SFRs.

MOV A, @R0; Means: Store the value from the register whose address is in the R0 register into accumulator.

On indirect addressing, the registers R0, R1 or Stack Pointer are used for specifying 8-bit addresses. Since only 8 bits are available, it is possible to access only registers of internal RAM in this way (128 locations in former or 256 locations in latest versions of the microcontrollers). If memory extension in form of additional memory chip is used then the 16-bit DPTR Register (consisting of the registers DPTRL and DPTRH) is used for specifying addresses. In this way, it is possible to access any location in the range of 64K.

3.8 EXTERNAL MEMORY

3.8.1 External Program Memory

External Program Memory and external Data Memory may be combined if desired by applying the RD and PSEN signals to the inputs of an AND gate and using the output of the gate as the read strobe to the external Program/Data memory.

The hardware configuration for external program execution is shown in Fig. 3.23. Note that 16 I/O lines (Ports 0 and 2) are dedicated to bus functions during external Program Memory fetches. Port 0 (P0 in Fig. 3.23) serves as a multiplexed address/data bus. It emits the low byte of the Program Counter (PCL) as an address, and then goes into a float state awaiting the arrival of the code byte from the Program Memory. During the time that the low byte of the Program Counter is valid on P0, the signal ALE (Address Latch Enable) clocks this byte into an address latch. Meanwhile, Port 2 (P2 in Fig. 3.23) emits the high byte of the Program Counter (PCH). Then PSEN strobes the EPROM and the code byte is read into the microcontroller.

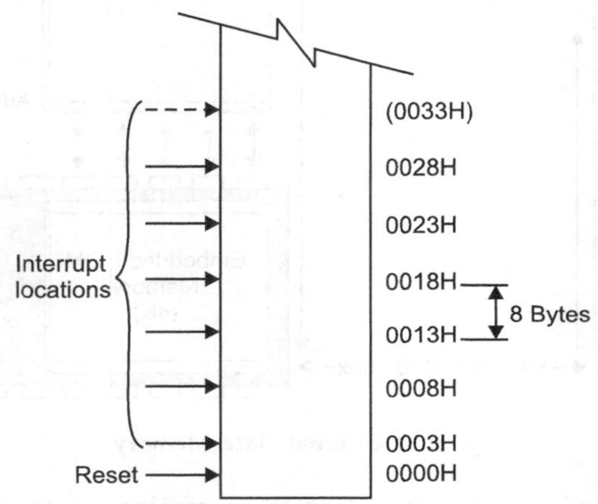

Fig. 3.22 External Program Memory

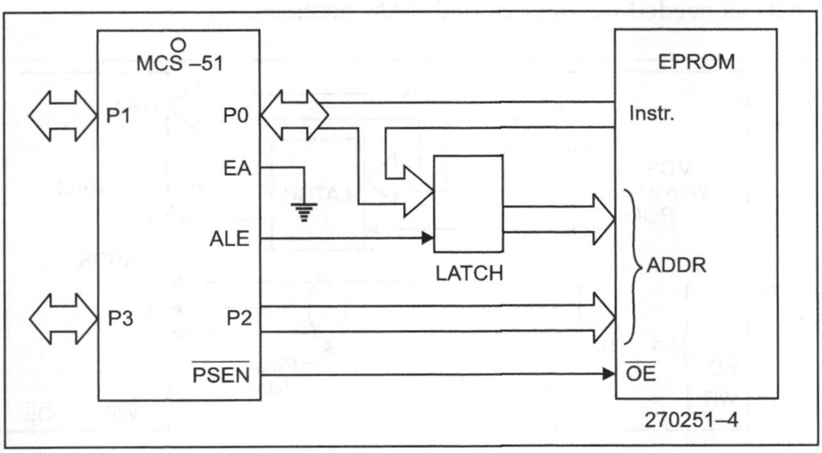

Fig. 3.23 Executing from External

Program Memory

Program Memory addresses are always 16 bits wide, even though the actual amount of Program Memory used may be less than 64K bytes. External program execution sacrifices two of the 8-bit ports, P0 and P2, to the function of addressing the Program Memory.

3.8.2 External Data Memory

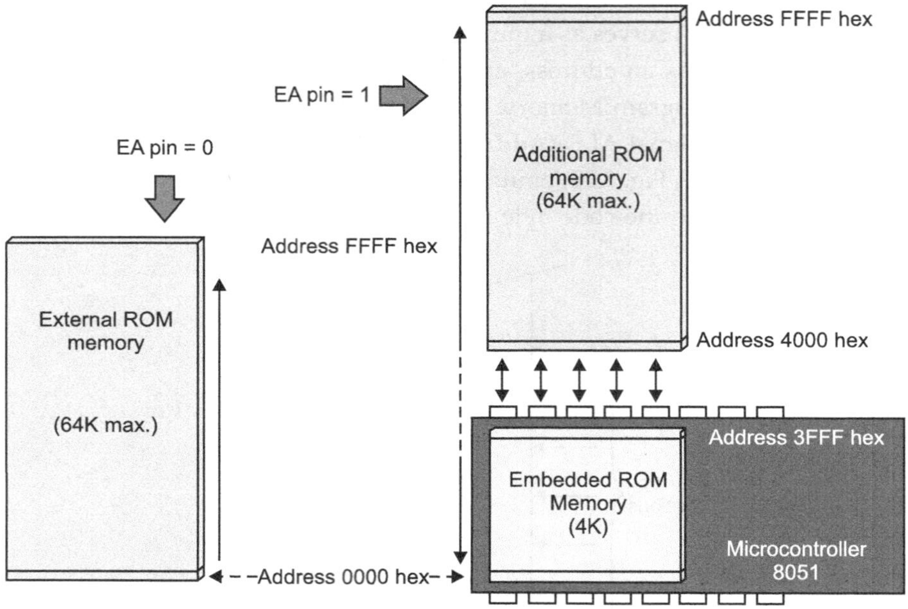

Fig. 3.24 External Data Memory

Figure 3.24 shows a hardware configuration for accessing up to 2K bytes of external RAM. The CPU in this case is executing from internal ROM. Port 0 serves as a multiplexed address/ data bus to the RAM, and 3 lines of Port 2 are being used to page the RAM. The CPU generates RD and WR signals as needed during external RAM accesses.

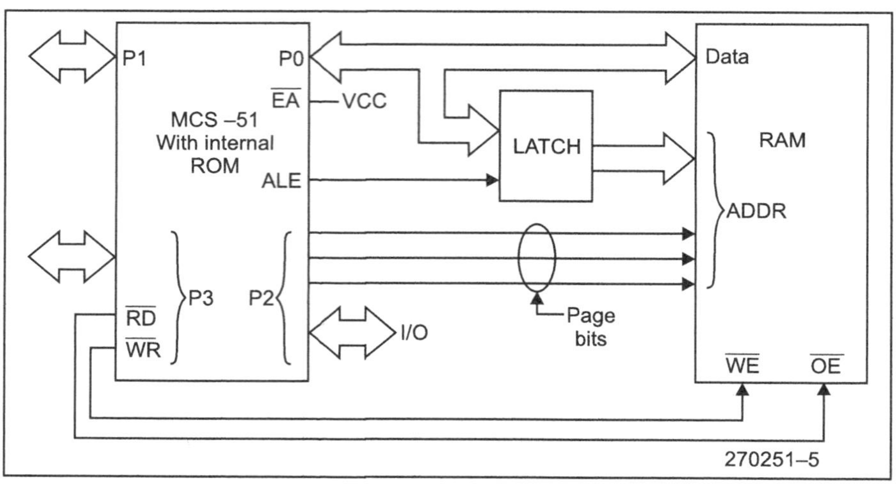

Fig. 3.25 Accessing External Data Memory

If the Program Memory is Internal, the other bits of P2 are available as I/O. There can be up to 64K bytes of external Data Memory. External Data Memory addresses can be either 1 or 2 bytes wide. One-byte addresses are often used in conjunction with one or more other I/O lines to page the RAM, as shown in Fig. 3.25. Two-byte addresses can also be used, in which case the high address byte is emitted at Port 2.

Counters and Timers

As explained in the previous chapter, the main oscillator of the microcontroller uses quartz crystal for its operating. As the frequency of this oscillator is precisely defined and very stable, these pulses are the most suitable for time measuring (such oscillators are used in quartz clocks as well). In order to measure time between two events it is only needed to count up pulses from this oscillator. That is exactly what the timer is doing. Namely, if the timer is properly programmed, the value written to the timer register will be incremented or decremented after each coming pulse, *i.e.*, once per each machine cycle cycle. Taking into account that one instruction lasts 12 quartz oscillator periods (one machine cycle), by embedding quartz with oscillator frequency of 12MHz, a number in the timer register will be changed million times per second, *i.e.*, each microsecond.

The 8051 microcontrollers have 2 timer counters called T0 and T1. As their names tell, their main purposes are to measure time and count external events. Besides, they can be used for generating clock pulses used in serial communication, *i.e.*, Baud Rate.

Timer T0

As it is shown in the picture below, this timer consists of two registers – TH0 and TL0. The numbers these registers include represent a lower and a higher byte of one 16-digit binary number.

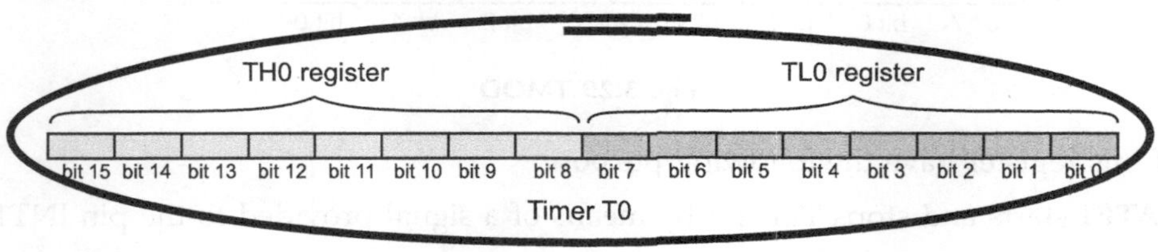

Fig. 3.26 Timer T0

This means that if the content of the timer 0 is equal to 0 (T0 = 0), then both registers it includes will include 0. If the same timer contains for example, number 1000 (decimal), then the register TH0 (higher byte) will contain number 3, while TL0 (lower byte) will contain decimal number 232.

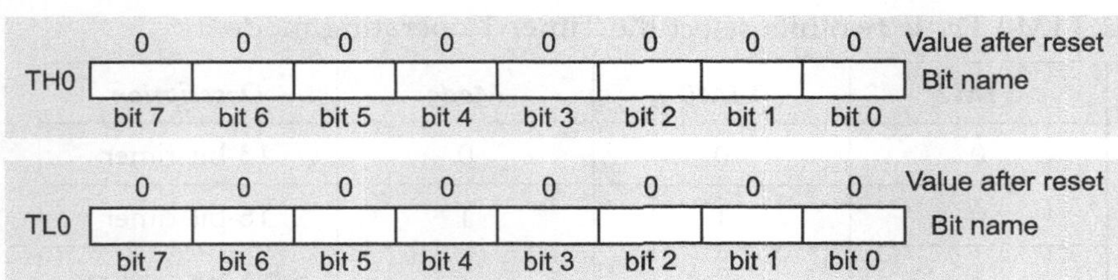

Fig. 3.27 Timer Lower Byte and Higher Byte

Formula used to calculate values in registers is very simple:

$$TH0 \times 256 + TL0 = T$$

Matching the previous example it would be as follows:

$$3 \times 256 + 232 = 1000$$

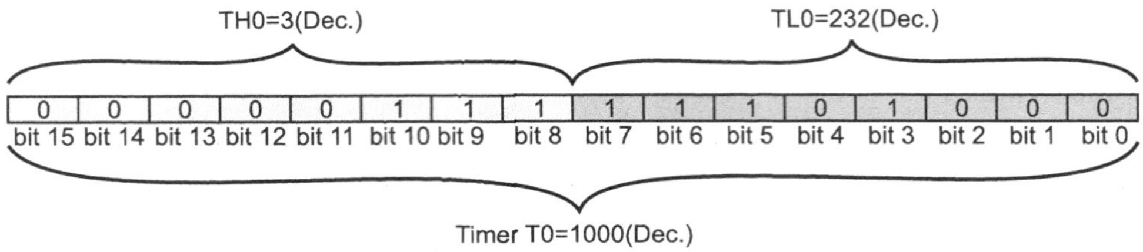

Fig. 3.28 Formula used in Timer 0

Since, the timers are virtually 16-bit registers, the greatest value that could be written to them is 65 535. In case of exceeding this value, the timer will be automatically reset and afterwards that counting starts from 0. It is called overflow. Two registers TMOD and TCON are closely connected to this timer and control how it operates.

TMOD Register (Timer Mode)

This register selects mode of the timers T0 and T1. As illustrated in the following picture, the lower 4 bits (bit0–bit3) refer to the timer 0, while the higher 4 bits (bit4–bit7) refer to the timer 1. There are in total of 4 modes and each of them is described here in this book.

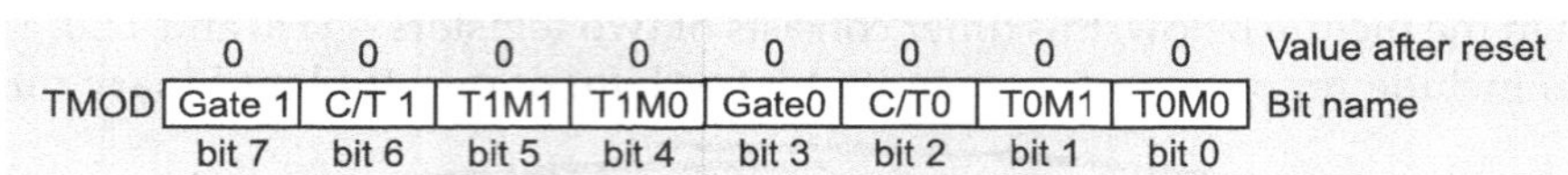

Fig. 3.29 TMOD

Bits of this register have the following purpose:

- **GATE1** starts and stops Timer 1 by means of a signal provided to the pin INT1 (P3.3):
  - **1** - Timer 1 operates only if the bit INT1 is set
  - **0** - Timer 1 operates regardless of the state of the bit INT 1
- **C/T1** selects which pulses are to be counted up by the timer/counter 1:
  - **1** - Timer counts pulses provided to the pin T1 (P3.5)
  - **0** - Timer counts pulses from internal oscillator
- **T1M1,T1M0** These two bits select the Timer 1 operating mode.

T1M1	*T1M0*	*Mode*	*Description*
0	0	0	13-bit timer
0	1	1	16-bit timer
1	0	2	8-bit auto-reload
1	1	3	Split mode

- **GATE0** starts and stops Timer 1, using a signal provided to the pin INT0 (P3.2):
  - **1** - Timer 0 operates only if the bit INT0 is set
  - **0** - Timer 0 operates regardless of the state of the bit INT0

- **C/T0** selects which pulses are to be counted up by the timer/counter 0:
  - **1** - Timer counts pulses provided to the pin T0(P3.4)
  - **0** - Timer counts pulses from internal oscillator
- **T0M1,T0M0** These two bits select the Timer 0 operating mode.

T0M1	*T0M0*	*Mode*	*Description*
0	0	0	13-bit timer
0	1	1	16-bit timer
1	0	2	8-bit auto-reload
1	1	3	Split mode

Timer 0 in mode 0 (13-bit timer)

This is one of the rarities being kept only for compatibility with the previous versions of the microcontrollers. When using this mode, the higher byte TH0 and only the first 5 bits of the lower byte TL0 are in use. Being configured in this way, the Timer 0 uses only 13 of all 16 bits. How does it operate? With each new pulse coming, the state of the lower register (that one with 5 bits) is changed. After 32 pulses received it becomes full and automatically is reset, while the higher byte TH0 is incremented by 1. This action will be repeated until registers count up 8192 pulses. After that, both registers are reset and counting starts from 0.

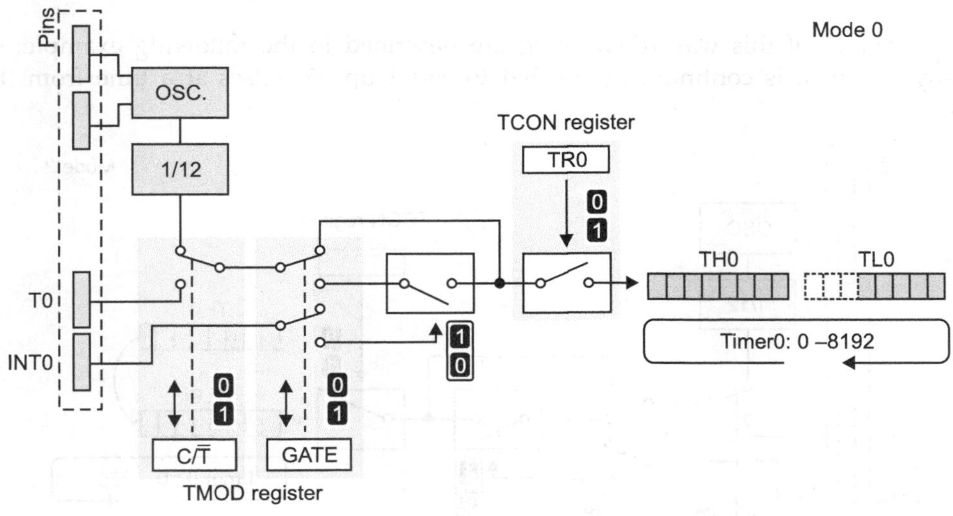

Fig. 3.30 TMOD Register

Timer 0 in mode 1 (16-bit timer)

All bits from the registers TH0 and TL0 are used in this mode. That is why for this mode is being more commonly used. Counting is performed in the same way as in mode 0, with difference that the timer counts up to 65 536, *i.e.,* as far as the use of 16 bits allows.

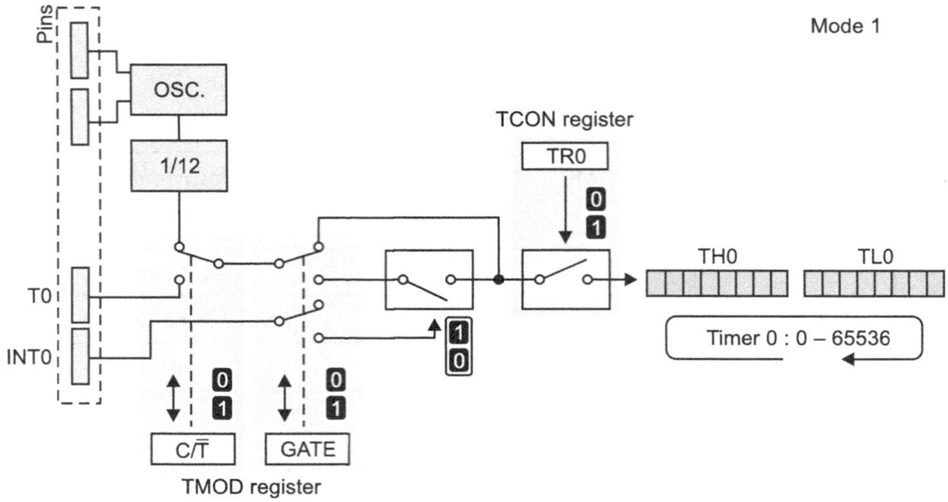

Fig. 3.31 Timer Mode 1

Timer 0 in mode 2 (Auto-Reload Timer)

What does auto-reload mean? Simply, it means that such timer uses only one 8-bit register for counting, but it never counts from 0 but from an arbitrary chosen value (0 – 255) saved in another register.

The advantages of this way of counting are described in the following example: suppose that for any reason it is continuously needed to count up 55 pulses at a time from the clock generator.

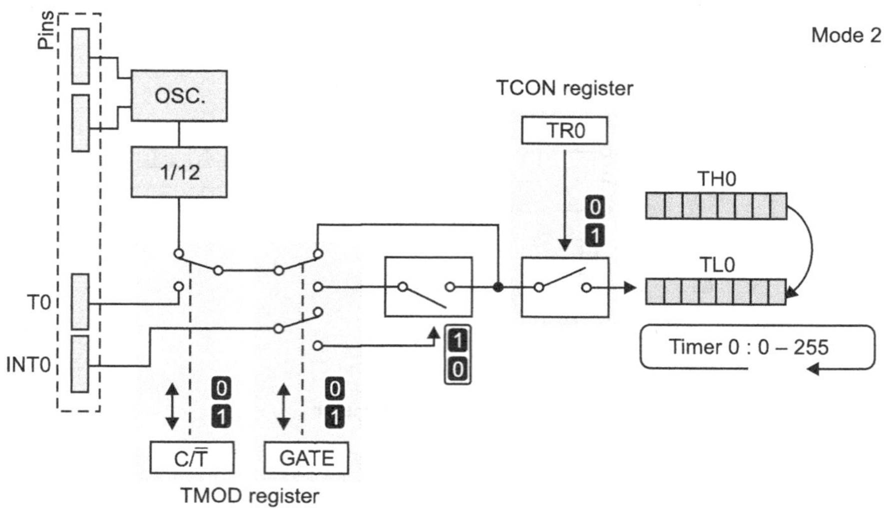

Fig. 3.32 Timer 0 in Mode 2

When using mode 1 or mode 0, it is needed to write number 200 to the timer registers and check constantly afterwards whether overflow occurred, *i.e.*, whether the value 255 is reached by counting. When it has occurred, it is needed to rewrite number 200 and repeat the whole procedure. The microcontroller performs the same procedure in mode 2 automatically. Namely, in this mode it is only register TL0 operating as a timer (normally 8-bit), while the value from which counting should start is saved in the TH0 register. Referring to the previous example, in order to register each 55th pulse, it is needed to write the number 200 to the register and configure the timer to operate in mode 2.

Timer 0 in Mode 3 (Split Timer)

By configuring Timer 0 to operate in Mode 3, the 16-bit counter consisting of two registers TH0 and TL0 is split into two independent 8-bit timers. In addition, all control bits which belonged to the initial Timer 1 (consisting of the registers TH1 and TL1), now control newly created Timer 1. This means that even though the initial Timer 1 still can be configured to operate in any mode (mode 1, 2 or 3), it is no longer able to stop, simply because there is no bit to do that. Therefore, in this mode, it will uninterruptedly "operate in the background."

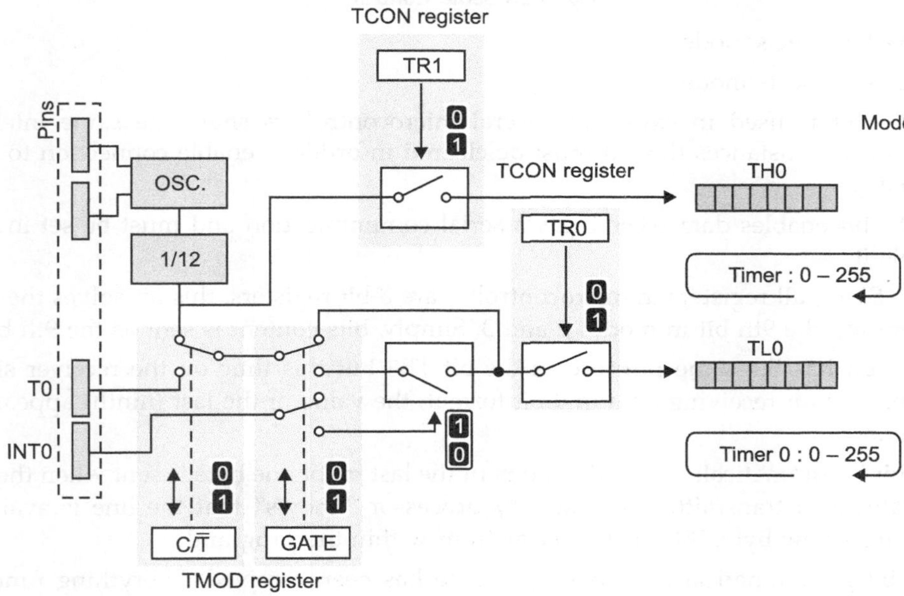

Fig. 3.33 Timer 0 in Mode 3

The only application of this mode is in case two independent 'quick' timers are used and the initial Timer 1 whose operating is out of control is used as baud rate generator.

3.9 UART (UNIVERSAL ASYNCHRONOUS RECEIVER AND TRANSMITTER)

One of the features that make this microcontroller so powerful is an integrated UART, better known as a serial port. It is a duplex port, which means that it can transmit and receive data simultaneously. Without it, serial data sending and receiving would be endlessly complicated part of the program where the pin state continuously is being changed and checked according to strictly determined rhythm. Naturally, it does not happen here because the UART resolves it in a very elegant manner. All the programmer needs to do is to simply select serial port mode and baud rate. When the programmer is such configured, serial data sending is done by writing to the register SBUF while data receiving is done by reading the same register. The microcontroller takes care of all issues necessary for not making any error during data exchange.

	bit7	bit6	bit5	bit4	bit3	bit2	bit1	bit0	
	X	X	X	X	X	X	X	X	Value after reset
SBUF									Bit name

Fig. 3.34 Serial Buffer

Serial port should be configured prior to being used. That determines how many bits one serial "word" contains, what the baud rate is and what the pulse source for synchronization is. After controlling this all bits are stored in the SFR Register SCON (Serial Control).

SCON Register (Serial Port Control Register)

	bit7	bit6	bit5	bit4	bit3	bit2	bit1	bit0	
	0	0	0	0	0	0	0	0	Value after reset
SCON	SM0	SM1	SM2	REN	TB8	RB8	TI	RI	Bit name

Fig. 3.35 Serial Control

- **SM0** - bit selects mode
- **SM1** - bit selects mode
- **SM2** - bit is used in case that several microcontrollers share the same interface. In normal circumstances this bit must be cleared in order to enable connection to function normally.
- **REN** - bit enables data receiving via serial communication and must be set in order to enable it.
- **TB8** - Since, all registers in microcontroller are 8-bit registers, this bit solves the problem of sending the 9th bit in modes 2 and 3. Simply, bits content is sent as the 9th bit.
- **RB8** - bit has the same purpose as the bit TB8 but this time on the receiver side. This means that on receiving data in 9-bit format, the value of the last (ninth) appears on its location.
- **TI** - bit is automatically set at the moment the last bit of one byte is sent when the USART operates as a transmitter. In that way processor "knows" that the line is available for sending a new byte. Bit must be clear from within the program!
- **RI** - bit is automatically set once one byte has been received. Everything functions in the similar way as in the previous case but on the receive side. This is line a "doorbell" which announces that a byte has been received via serial communication. It should be read quickly prior to a new data takes its place. This bit must also be also cleared from within the program!

As seen, serial port mode is selected by combining the bits SM0 and SM2:

SM0	SM1	Mode	Description	Baud Rate
0	0	0	8-bit Shift Register	1/12 the quartz frequency
0	1	1	8-bit UART	Determined by the timer 1
1	0	2	9-bit UART	1/32 the quartz frequency (1/64 the quartz frequency)
1	1	3	9-bit UART	Determined by the timer 1

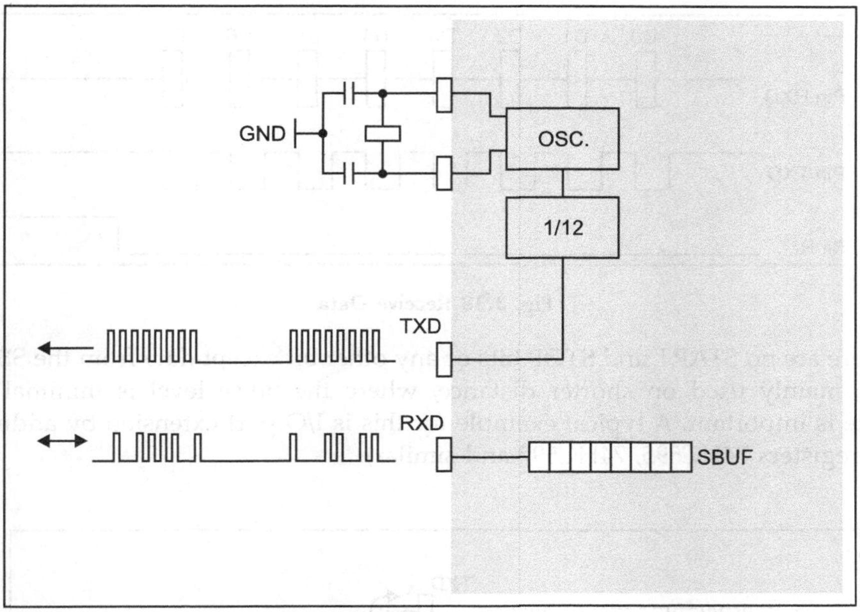

Fig. 3.36 Serial Buffer

In mode 0, the data are transferred through the RXD pin, while clock pulses appear on the TXD pin. The bout rate is fixed at 1/12 the quartz oscillator frequency. On transmit; the least significant bit (LSB bit) is being sent/received first. (Received).

TRANSMIT - Data transmission in form of pulse train automatically starts on the pin RXD at the moment the data has been written to the SBUF register. In fact, this process starts after any instruction being performed on this register. Upon all 8 bits have been sent, the bit TI in the SCON register is automatically set.

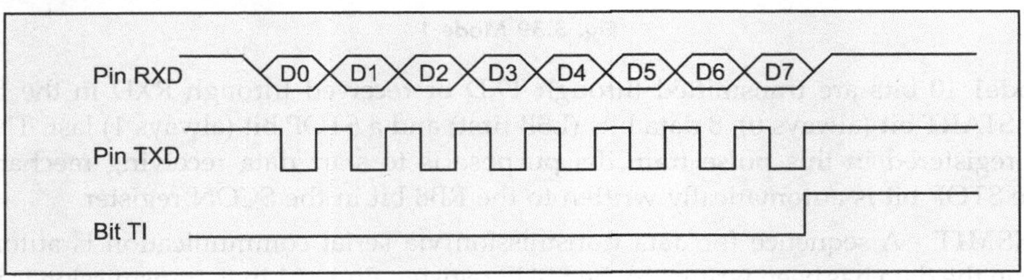

Fig. 3.37 Transmit Data

RECEIVE - Starts data receiving through the pin RXD once two necessary conditions are met: bit REN=1 and RI=0 (both bits reside in the SCON register). Upon 8 bits have been received, the bit RI (register SCON) is automatically set, which indicates that one byte is received.

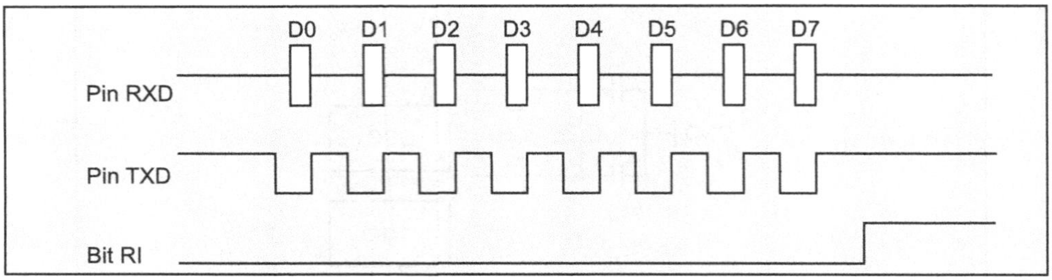

Fig. 3.38 Receive Data

Since, there are no START and STOP bits or any other bit except data from the SBUF register, this mode is mainly used on shorter distance, where the noise level is minimal and where operating rate is important. A typical example for this is I/O port extension by adding cheap IC circuit (shift registers 74HC595, 74HC597 and similar).

Mode 1

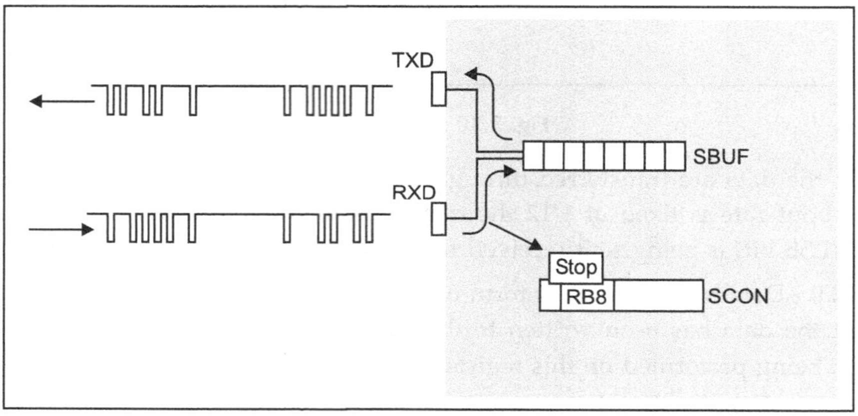

Fig. 3.39 Mode 1

In Mode1 10 bits are transmitted through TXD or received through RXD in the following manner: a START bit (always 0), 8 data bits (LSB first) and a STOP bit (always 1) last. The START bit is not registered in this pulse train. Its purpose is to start data receiving mechanism. On receive the STOP bit is automatically written to the RB8 bit in the SCON register.

TRANSMIT - A sequence for data transmission via serial communication is automatically started upon the data has been written to the SBUF register. End of 1 byte transmission is indicated by setting the TI bit in the SCON register.

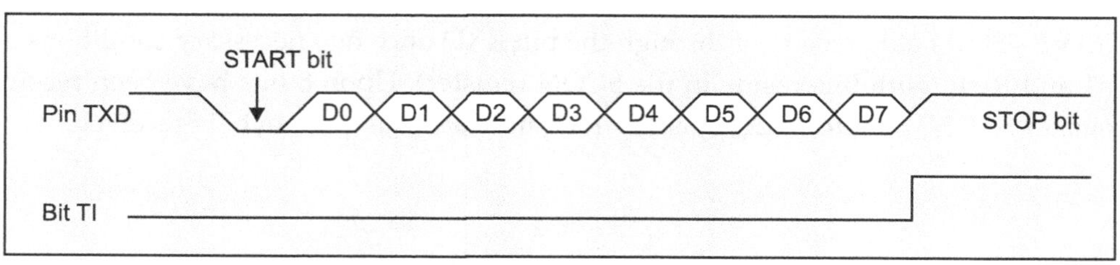

Fig. 3.40 Transmit Data

RECEIVE - Receiving starts as soon as the START bit (logic zero (0)) appears on the pin RXD. The condition is that bit REN=1and bit RI=0. Both of them are stored in the SCON register. The RI bit is automatically set upon receiving has been completed.

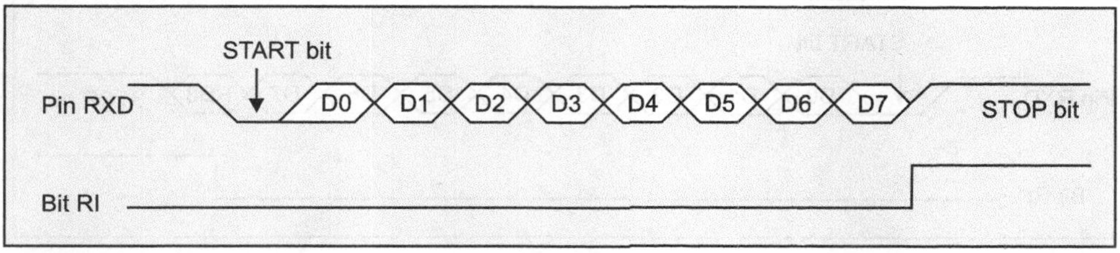

Fig. 3.41 Receive Data

The Baud rate in this mode is determined by the timer 1 overflow time.

Mode 2

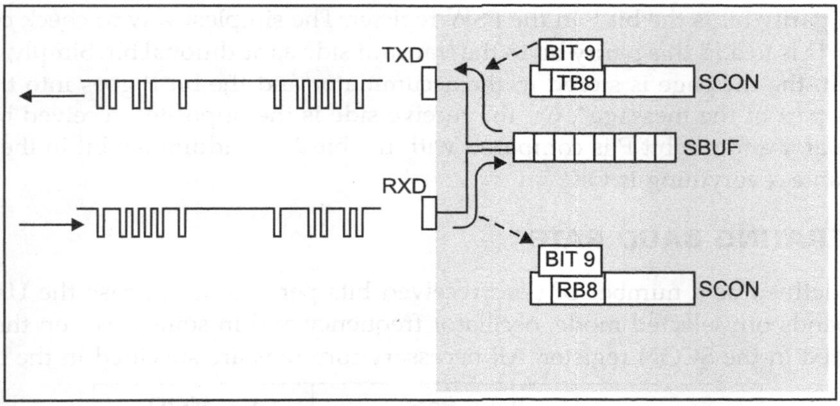

Fig. 3.42 Mode 2 Transmit and Receive Data

In mode 2, 11 bits are sent through TXD or received through RXD: a START bit (always 0), 8 data bits (LSB first), additional 9th data bit and a STOP bit (always 1) last. On transmit; the 9th data bit is actually the TB8 bit from the SCON register. This bit commonly has the purpose of parity bit. Upon transmission, the 9th data bit is copied to the RB8 bit in the same register (SCON).The baud rate is either 1/32 or 1/64 the quartz oscillator frequency.

TRANSMIT - A sequence for data transmission via serial communication is automatically started upon the data has been written to the SBUF register. End of 1 byte transmission is indicated by setting the TI bit in the SCON register.

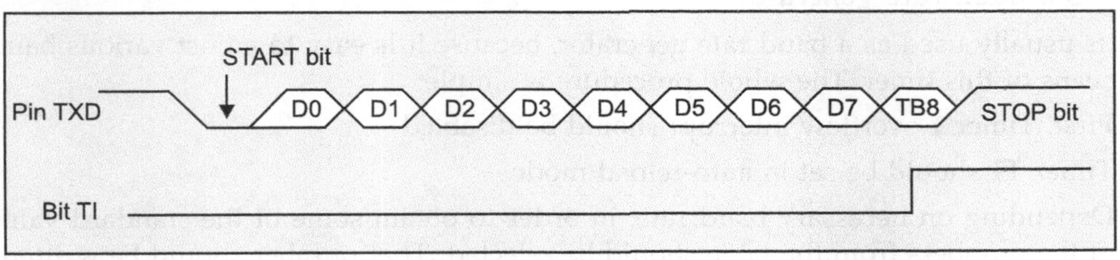

Fig. 3.43 Transmit Data

RECEIVE - Receiving starts as soon as the START bit (logic zero (0)) appears on the pin RXD. The condition is that bit REN=1and bit RI=0. Both of them are stored in the SCON register. The RI bit is automatically set upon receiving has been completed.

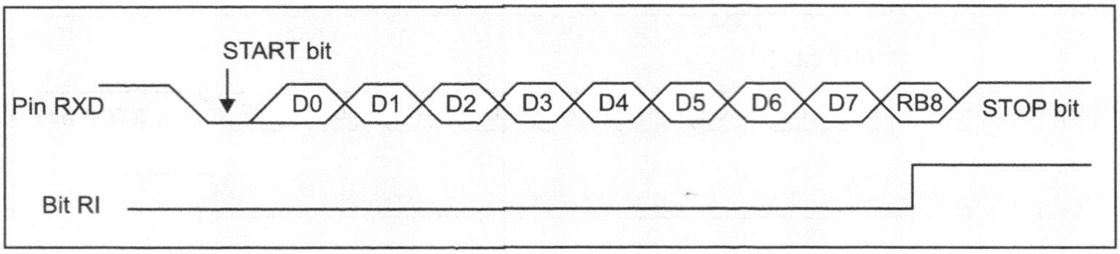

Fig. 3.44 Receive Data

Mode 3

Mode 3 is the same as Mode 2 except the baud rate. In Mode 3 is variable and can be selected.

Note: The parity bit is the bit P in the PSW register. The simplest way to check correctness of the received byte is to add this parity bit to the transmit side as additional bit. Simply, immediately before transmit, the message is stored in the accumulator and the bit P goes into the TB8 bit in order to be "a part of the message". On the receive side is the opposite : received byte is stored in the accumulator and the bit P is compared with the bit RB8 (additional bit in the message). If they are the same- everything is OK!

3.10 GENERATING BAUD RATE

Baud Rate is defined as a number of send/received bits per second. In case the UART is used, baud rate depends on: selected mode, oscillator frequency and in some cases on the state of the bit SMOD stored in the SCON register. All necessary formulas are specified in the table:

	Baud Rate	BitSMOD
Mode 0	Fosc. / 12	
Mode 1	1 Fosc. 16 12 (256-TH1)	BitSMOD
Mode 2	Fosc. / 32 Fosc. / 64	1 0
Mode 3	1 Fosc. 16 12 (256-TH1)	

Timer 1 as a baud rate generator

Timer 1 is usually used as a baud rate generator, because it is easy to adjust various baud rates by the means of this timer. The whole procedure is simple:

- First, Timer 1 overflow interrupt should be disabled
- Timer T1 should be set in auto-reload mode
- Depending on necessary baud rate, in order to obtain some of the standard values one of the numbers from the table should be selected. That number should be written to the TH1 register. That's all.

Baud Rate	Fosc. (MHz)					Bit SMOD
	11.0592	12	14.7456	16	20	
150	40 h	30 h	00 h			0
300	A0 h	98 h	80 h	75 h	52 h	0
600	D0 h	CC h	C0 h	BB h	A9 h	0
1200	E8 h	E6 h	E0 h	DE h	D5 h	0
2400	F4 h	F3 h	F0 h	EF h	EA h	0
4800		F3 h	EF h	EF h		1
4800	FA h		F8 h		F5 h	0
9600	FD h		FC h			0
9600					F5 h	1
19200	FD h		FC h			1
38400			FE h			1
76800			FF h			1

3.11 THE POWER MODE CONTROL (PCON) SPECIAL FUNCTION REGISTER

Conditionally said microcontroller is the most part of its "lifetime" is inactive for some external signal in order to takes its role in a show. It can make a great problem in case batteries are used for power supply. In extremely cases, the only solution is to put the whole electronics to sleep in order to reduce consumption to the minimum. A typical example of this is remote TV controller: it can be out of use for months but when used again it takes less than a second to send a command to TV receiver. While normally operating, the AT89S53 uses current of approximately 25 mA, which shows that it is not too sparing microcontroller. Anyway, it doesn't have to be always like this, it can easily switch the operation mode in order to reduce its total consumption to approximately 40uA. Actually, there are two power-saving modes of operation: *Idle* and *Power Down.*

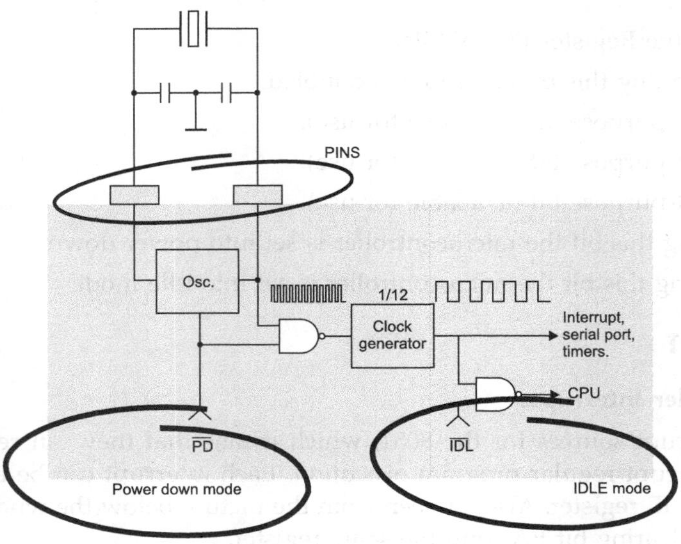

Fig. 3.45 PCON

Idle Mode

Immediately upon instruction which sets the bit IDL in the PCON register, the microcontroller turns off the greatest power consumer- CPU unit while peripheral unit's serial port, timers and interrupt system continue operating normally consuming 6.5mA. In Idle mode, the state of all registers and I/O ports remains unchanged.

In order to terminate the Idle mode and make the microcontroller operate normally, it is necessary to enable and execute any interrupt or reset. Then, the IDL bit is automatically cleared and the program continues executing from instruction following that instruction which has set the IDL bit. It is recommended that three first following one which set NOP instructions. They do not perform any operation but keep the microcontroller from undesired changes on the I/O ports.

Power Down Mode

When the bit PD in the register PCON is set from within the program, the microcontroller is set to Power down mode. It and turns off its internal oscillator reducing drastically consumption in that way. In power- down mode the microcontroller can operate using only 2V power supply while the total power consumption is less than 40uA. The only way to get the microcontroller back to normal mode is reset.

During Power Down mode, the state of all SFR registers and I/O ports remains unchanged, and after the microcontroller is put get into the normal mode, the content of the SFR register is lost, but the content of internal RAM is saved. Reset signal must be long enough approximately 10 mS in order to stabilize quartz oscillator operating.

PCON									Register
	0	X	X	X	0	0	0	0	Value after reset
PCON	SMOD				GF1	GF0	PD	IDL	Bit name
	bit7	bit6	bit5	bit4	bit3	bit2	bit1	bit0	

Fig. 3.46 Power Control

The purpose of the Register PCON bits:

- SMOD By setting this bit baud rate is doubled.
- GF1 General-purpose bit (available for use).
- GF1 General-purpose bit (available for use).
- GF0 General-purpose bit (available for use).
- PD by setting this bit the microcontroller is set into power down
- IDL by setting this bit the microcontroller is set into idle mode

3.12 INTERRUPT

8051 Microcontroller Interrupts

There are five interrupt sources for the 8051, which means that they can recognize 5 different events that can interrupt regular program execution. Each interrupt can be enabled or disabled by setting bits in the IE register. Also, as seen from the picture below the whole interrupt system can be disabled by clearing bit EA from the same register.

Now, one detail should be explained which is not completely obvious but refers to external interrupts- INT0 and INT1. Namely, if the bits IT0 and IT1 stored in the TCON register are set, program interrupt will occur on changing logic state from 1 to 0, (only at the moment). If these bits are cleared, the same signal will generate interrupt request and it will be continuously executed as far as the pins are held low.

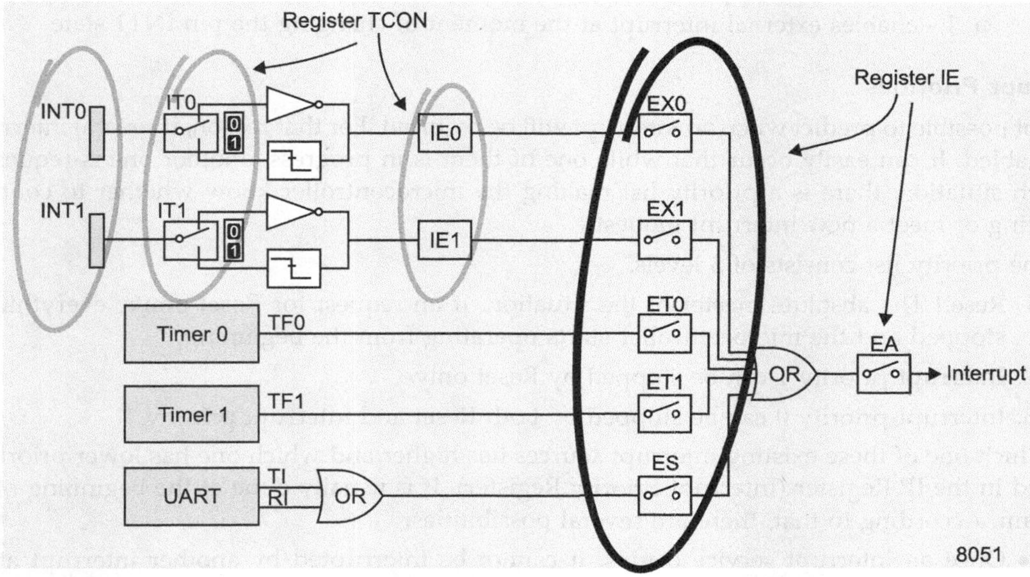

Fig. 3.47 8051 Microcontroller Interrupt

IE Register (Interrupt Enable)

	bit7	bit6	bit5	bit4	bit3	bit2	bit1	bit0
Value after reset	X	X	0	0	0	0	0	0
IP (Bit name)			PT2	PS	PT1	PX1	PT0	PX0

Fig. 3.48 Interrupt Enable

- **EA** - bit enables or disables all other interrupt sources (globally)
  - 0 - (when cleared) any interrupt request is ignored (even if it is enabled)
  - 1 - (when set to 1) enables all interrupts requests which are individually enabled
- **ES** - bit enables or disables serial communication interrupt (UART)
  - 0 - UART System cannot generate interrupt
  - 1 - UART System enables interrupt
- **ET1** - bit enables or disables Timer 1 interrupt
  - 0 - Timer 1 cannot generate interrupt
  - 1 - Timer 1 enables interrupt
- **EX1** - bit enables or disables INT 0 pin external interrupt
  - 0 - change of the pin INT0 logic state cannot generate interrupt
  - 1 - enables external interrupt at the moment of changing the pin INT0 state
- **ET0** - bit enables or disables timer 0 interrupt

  - 0 - Timer 0 cannot generate interrupt
  - 1 - enables timer 0 interrupt
- **EX0** - bit enables or disables INT1 pin external interrupt
  - 0 - change of the INT1 pin logic state cannot cause interrupt
  - 1 - enables external interrupt at the moment of changing the pin INT1 state

Interrupt Priorities

It is not possible to predict when an interrupt will be required. For that reason, if several interrupts are enabled. It can easily occur that while one of them is in progress, another one is requested. In such situation, there is a priority list making the microcontroller know whether to continue operating or meet a new interrupt request.

The priority list consists of 3 levels:

1. Reset! The absolute master of the situation. If an request for Reset omits, everything is stopped and the microcontroller starts operating from the beginning.
2. Interrupt priority 1 can be stopped by Reset only.
3. Interrupt priority 0 can be stopped by both Reset and interrupt priority 1.

Which one of these existing interrupt sources has higher and which one has lower priority is defined in the IP Register (Interrupt Priority Register). It is usually done at the beginning of the program. According to that, there are several possibilities:

- Once an interrupt service begins. It cannot be interrupted by another interrupt at the same or lower priority level, but only by a higher priority interrupt.
- If two interrupt requests, at different priority levels, arrive at the same time then the higher priority interrupt is serviced first.
- If the both interrupt requests, at the same priority level, occur one after another, the one who came later has to wait until routine being in progress ends.
- If two interrupts of equal priority requests arrive at the same time then the interrupt to be serviced is selected according to the following priority list :

1. External interrupt INT0
2. Timer 0 interrupt
3. External Interrupt INT1
4. Timer 1 interrupt
5. Serial Communication Interrupt

IP Register (Interrupt Priority)

The IP register bits specify the priority level of each interrupt (high or low priority).

	bit7	bit6	bit5	bit4	bit3	bit2	bit1	bit0	
	X	X	0	0	0	0	0	0	Value after reset
IP			PT2	PS	PT1	PX1	PT0	PX0	Bit name

Fig. 3.49 Interrupt Priority

- **PS** - Serial Port Interrupt priority bit
  - Priority 0

  - Priority 1
- **PT1** - Timer 1 interrupt priority
  - Priority 0
  - Priority 1
- **PX1** - External Interrupt INT1 priority
  - Priority 0
  - Priority 1
- **PT0** - Timer 0 Interrupt Priority
  - Priority 0
  - Priority 1
- **PX0** - External Interrupt INT0 Priority
  - Priority 0
  - Priority 1

Handling Interrupt

Once some of interrupt requests arrives, everything occurs according to the following order:

1. Instruction in progress is ended
2. The address of the next instruction to execute is pushed on the stack
3. Depending on which interrupt is requested, one of 5 vectors (addresses) is written to the program counter in accordance to the following table:

Interrupt Source	Vector (address)
IE0	3 h
TF0	B h
TF1	1B h
RI, TI	23 h
All addresses are in hexadecimal format	

4. The appropriate subroutines processing interrupts should be located at these addresses. Instead of them, there are usually jump instructions indicating the location where the subroutines reside.
5. When interrupt routine is executed, the address of the next instruction to execute is poped from the stack to the program counter and interrupted program continues operating from where it left off.

Interrupt Structure

The 8051 core provides 5 interrupt sources: 2 external interrupts, 2 timer interrupts and the serial port interrupt. What follows is an overview of the interrupt structure for the 8051. Other MCS-51 devices have additional interrupt sources and vectors as shown in Table 1. Refer to the appropriate chapters on other devices for further information on their interrupts.

Interrupt Enables

Each of the interrupt sources can be individually enabled or disabled by setting or clearing a bit in the SFR named IE (Interrupt Enable). This register also contains a global disable bit, which can be cleared to disable all interrupts at once. Figure 3.50 shows the IE register for the 8051.

(MSB) (LSB)

EA	—	—	ES	ET1	EX1	ET0	EX0

Enable bit = 1 enables the interrupt
Enable bit = 0 disables it

Symbol	Position	Function
EA	IE.7	disables all interrupts. If EA = 0, no interrupt will be acknowledged. If EA = 1, each interrupt source is individually enabled or disabled by setting or clearing its enable bit.
—	IE.6	reserved*
—	IE.5	reserved*
ES	IE.4	Serial Port interrupt enable bit
ET1	IE.3	Timer 1 Overflow interrupt enable bit
EX1	IE.2	External interrupt 1 enable bit
ET0	IE.1	Timer 0 Overflow interrupt enable bit
EX0	IE.0	External interrupt 0 enable bit

*These reserved bits are used in other MCS-51 devices.

Fig. 3.50 IE(Interrupt Enable) Register in the 8051

Interrupt Priorities

Each interrupt source can also be individually programmed to one of two priority levels by setting or clearing a bit in the SFR named IP (Interrupt Priority). Figure 3.51 shows the IP register in the 8051. A low-priority interrupt can be interrupted by a high priority interrupt, but not by another low-priority interrupt. A high-priority interrupt can't be interrupted by any other interrupt source. If two interrupt requests of different priority levels are received simultaneously, the request of higher priority level is serviced. If interrupt requests of the same priority level are received simultaneously, an internal polling sequence determines which request is serviced. Thus within each priority level there is a second priority structure determined by the polling sequence. Figure 19 shows, for the 8051, how the IE and IP registers and the polling sequence work to determine which if any interrupt will be serviced.

(MSB) (LSB)

—	—	—	PS	PT1	PX1	PT0	PX0

Priority bit = 1 assigns high priority
Priority bit = 0 assigns low priority

Symbol	Position	Function
—	IP.7	reserved*
—	IP.6	reserved*
—	IP.5	reserved*
PS	IP.4	Serial Port interrupt priority bit
PT1	IP.3	Timer 1 interrupt priority bit
PX1	IP.2	External interrupt 1 priority bit
PT0	IP.1	Timer 0 interrupt priority bit
PX0	IP.0	External interrupt 0 priority bit

*These reserved bits are used in other MCS-51 devices.

Fig. 3.51 IP(Interrupt Priority) Register in the 8051

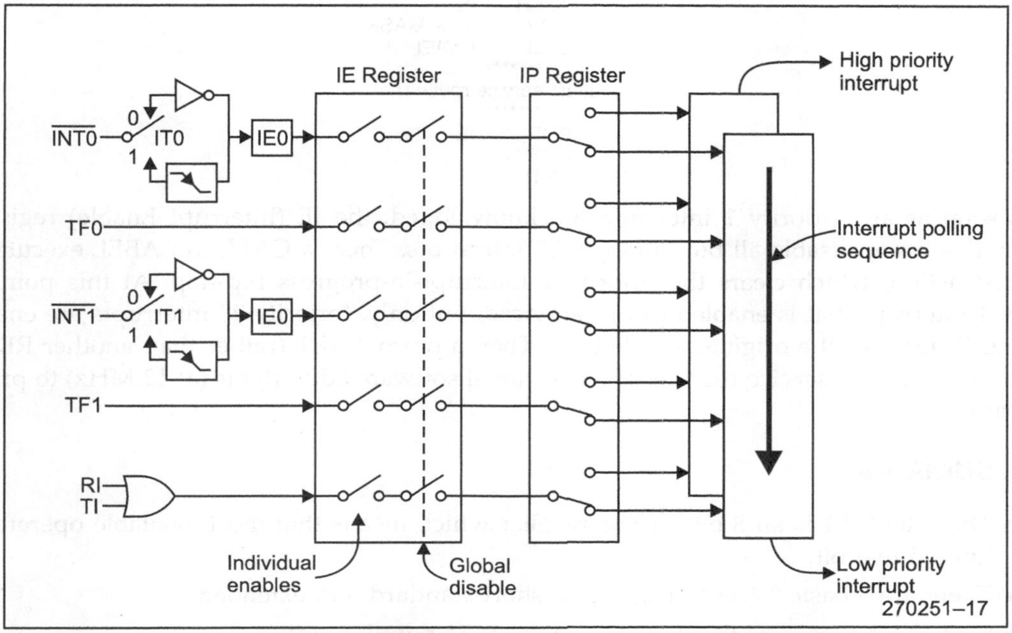

Fig. 3.52 8051 Interrupt Control System

In operation, all the interrupt flags are latched into the interrupt control system during State 5 of every machine cycle. The samples are polled during the following machine cycle. If the flag for an enabled interrupt is found to be set (1), the interrupt system generates an LCALL to the appropriate location in Program Memory, unless some other condition blocks the interrupt. Several conditions can block an interrupt, among them that an interrupt of equal or higher priority level is already in progress. The hardware-generated LCALL causes the contents of the Program Counter to be pushed onto the stack, and reloads the PC with the beginning address of the service routine. As previously noted (Figure 3), the service routine for each interrupt begins at a fixed location.

Only the Program Counter is automatically pushed onto the stack, not the PSW or any other register. Having only the PC be automatically saved allows the programmer to decide how much time to spend saving which other registers. This enhances the interrupt response time, albeit at the expense of increasing the programmer's burden of responsibility. As a result, many interrupt functions that are typical in control applications toggling a port pin, for example, or reloading a timer, or unloading a serial buffer scan often be completed in less time than it takes other architectures to commence them.

Simulating A Third Priority Level In Software

Some applications require more than the two priority levels that are provided by on-chip hardware in MCS-51 devices. In these cases, relatively simple software can be written to produce the same effect as a third priority level. First, interrupts that are to have higher priority than 1 are assigned to priority 1 in the IP (Interrupt Priority) register. The service routines for priority 1 interrupts that are supposed to be interruptible by "priority 2" interrupts are written to include the following code:

```
        PUSH   IE
        MOV    IE,* MASK
        CALL   LABEL
        *******
  (execute service routine)
        *******
        POP    IE
        RET
LABEL : RETI
```

As soon as any priority 1 interrupt is acknowledged, the IE (Interrupt Enable) register is re-defined so as to disable all but "priority 2" interrupts. Then, a CALL to LABEL executes the RETI instruction, which clears the priority 1 interrupt-in-progress flip-flop. At this point any priority 1 interrupt that is enabled can be serviced, but only "priority 2" interrupts are enabled. POPping IE restores the original enable byte. Then a normal RET (rather than another RETI) is used to terminate the service routine. The additional software adds 10 ms (at 12 MHz) to priority 1 interrupts.

3.13 SUMMARY

- The Intel 8051 is an 8 bit microcontroller which means that most available operator are limited to 8 bit.
- There are 3 basic "sizes" of the 8051: short standard and extended.
- Some of the features that have made the 8051 popular are
  - (*i*) 8 bit data bus
  - (*ii*) 16 data bus address bus
  - (*iii*) 34 general purpose registers each of 8 bits
  - (*iv*) 16 bit timers (usually 2,but may have more or less)
  - (*v*) 3 internal and 2 external interrupt.
  - (*vi*) Bit as well as byte addressable ram area AF 16 bytes.
  - (*vii*) Four 8-bit ports, short models have two 8-bit ports.
  - (*viii*) 16-bit program computer and data pointer.
- 8051 family chips make up over 50% of the embedded chip market.
- 8051 is a 40- pin dip IC.
- 8051 was introduced in 1980 by Intel corporation.
- A typical 8051 contain CPU with Boolean processor 5 or 6 interrupts:2 are extended 2 priority level Programmable full-duplex serial port (based rate provided by one if the timers) 32 input output lines (four 8-bit ports) Ram Rom/EPROM in some models.
- Serial port pins can "receive" current of 10ma.
- If all 8 bits on a port are achieve, total current must be limited to 71ma.
- The microcontroller memory is divided into program memory and data memory.
- All 8051 microcontrollers have 16-bit addressing bus and can address 64kb memory.
- Special function register(SFPS)are a kind of control table used for running and monitoring microcontrollers operating.
- Accumulator once an arithmetic operation is performed by the ALU, the result is placed into the accumulator.

- B register is used during multiply and divide operations.
- The program status word (PSW) contains several status bits that reflect the current state of the CPU.
- PSW register contains carry bit overflow flag parity bit and user definable status flag.
- DPTR (data pointer) consists of two separate registers DPH (data pointer high) and DPL (data pointer low).
- The DPTR register is usually used for strong data and intermediate results which have nothing location.
- The 8051 microcontrollers have 2 times counters called t_0 and t_1.
- Timer t_0 consists of two registers THO and TLO.
- TCON – timer control register.
- TMODE register – timer diode register.
- UART universal asynchronous receiver and transmitter.
- Icon register – serial port control register.
- Baud rate – is defined as a number of send/received bits per second.
- The priority bits consists of the situation if an register for reset omits, everything is stopped and the microcontrollers starts operating from the beginning.
- Interrupt priority 1 can be stopped by reset only.
- Interrupt priority 0 can be stopped by both reset and interrupt priority.
- PC (program counter) points to the address of the next instruction.
- Stack pointer (sp): the register used to access the stack is called stack pointer.

3.14 QUESTIONS

1. If a microcontroller has both 8 bit and 16 bit versions which of the following statements is true.
   (*a*) The 8 bit software will run on the 16 bit system.
   (*b*) The 4 bit software will run on the 16 bit system.
   (*c*) The 2 bit software will run on the 24 bit system.
   (*d*) The 8 bit software will run on the 8 bit system.
2. The 8051 has ______________ on-chip timers.
3. The 8051 microcontrollers has _________ pins for input output.
4. Registers $r_0 - r_7$ are all _____________ bits wide.
5. Name a 16 bit register in the 8051.
6. The PSW in a ________ bit register.
7. The 8751 on chip ROM is of type __________.
8. Internal data memory addresses are always _______byte.
9. The program memory can be up to _________bytes and data memory external to the chip.
10. On power of the 8051 uses bank ______for registers $r_0 - r_7$.

11. The 8051 dip package is a _____pin package.
12. ALE is an ________pin.
13. RST is an ________pin.
14. Find the machine cycle for the following crystal frequencies connected to x1 and x2 12 mhz 20 mhz 25 mhz 30 mhz.
15. PSEN stands for ________.
16. Which mode of the timer is used to set the band rate?
17. SCON stands for ________and it is a_________bit register.
18. Which timer of the 8051 is used to set the band rate?
19. To which register does the SMOD bit belong?
20. State its roll in the rate of data transfer.
21. The stack pointer in the 8051 microcontroller is a ______________.
22. The Intel 8051 microcontroller has the internal ROM of ______________ bytes and internal ram of ______________ bytes.
23. They are ______________ external interrupt in 8051.
24. The number of math flags in 8051 work?

    (*a*) 2 (*b*) 3 (*c*) 4 (*d*) 7
25. DPTR can address ________bytes.

    (*a*) 128 bytes (*b*) 1 kbytes (*c*) 64 bytes (*d*) 32 kbytes
26. The internal clock frequency of the microcontroller ranges from ________.

Chapter 4

MOVING DATA

4.1 INTRODUCTION

The microcontroller 8051 instructions set includes 110 instructions, 49 of which are single byte instructions, 45 are two bytes instructions and 17 are three bytes instructions. The instructions format consists of a function mnemonic followed by destination and source field.

All the instructions of microcontroller 8051 may be classified based on the functional aspect are given below

- Data transfer group.
- Arithmetic group.
- Logical group.
- Bit manipulation group.
- Branching or Control transfer group.

4.2 ADDRESSING MODES

The instructions of 8051 may be classified based on the source or destination type

- Register addressing.
- Direct addressing.
- Register Indirect addressing.
- Immediate addressing.
- Base register + Index register.

4.2.1 Immediate Addressing Modes

When the 8051 executes an immediate data move, the program counter is automatically incremented to point to the byte(s) following the opcode byte in the program memory. Whatever, data is found there is copied to the destination address. The mnemonic for immediate data is the pound sign (#).

MOV A, # 20H: Load 20H into A

MOV R2, #42H: Load the decimal value 42H into R2

MOV R0, # 24H: Load the decimal value 24H into R0

MOV DPTR, #4000H: DPTR=4000H

Although, the DPTR register is 16-bit, it can also be accessed as two 8-bit register, DPH & DPL where,

DPH is the high byte

DPL is the Low byte

That is MOV DPTR, # 4250H is the same as

MOV DPL, #50H

MOV DPH, # 42 H

Also, notice that the following would produce an error, since the value is larger that 16-bit

MOV DPTR, # 425000; illegal!! Values > 42000

4.2.2 Register Addressing Modes

Register addressing modes involves the use of register to hold the data to be manipulated. Example of register addressing mode follow.

```
MOV A, R!     ; Copy the contents of R1 into A
MOV R0, A     ; Copy contents of A into R2
ADD A, R5     ; Add the contents of R5 to contents
MOV R5, A     ; save accumulator of R5
```

4.2.3 Direct Addressing Modes

There are 128bytes of RAM is the 8051, The Internal RAM uses addressing from 00H to 7FH to address each byte.

- RAM locations 00-1FH are assigned to the from bands of eight working register R0 to R7
- RAM locations 20-2FH are set aside as bit addressable space to save single bit data
- RAM locations 30-7FH are available as a place to save byte sized data.

Example of Direct addressing mode

```
MOV R1, 20H      ; same content of RAM location 20H in R1
MOV 42H, A       ; same content of A is RAM location 42H
```

The SFR addresses between 80H to FFH, since the addresses 00H to 7FH are addresses of RAM memory inside the 8051. All the address spaces of 80 to FF are not used bytes the SFR, The unused locations 80H to FFH are reserved and must not be used by the 8051 programmer.

4.2.4 Indirect Addressing Mode

In the register indirect addressing mode, a register is used the contents of R0 and R1 often called a data pointer, as a pointer to location in 256 bytes block and the 256 bytes of internal RAM or the lower, 256 bytes of external data memory execution of PUSH and POP is also uses indirect addressing.

Example:

MOV @ R0, # n ; copy immediate byte A to the address in R0

MOV @ R0, A; copy A to RAM location RO pointers

MOV A, @ Rp ; copy the contents of the address in Rp to A.

Note:

Number in register Rp must be RAM address

R0 or R1 register for indirect addressing

Table opcode using immediate, register direct & indirect addressing mode

Mnemonic		**Operations**
MOV A, @ R0	→	Copy the content of the address in R0 to the A register
MOV @ R1, #35H	→	Copy the number 35H to the address in R1
MOV ADD,@R0	→	Copy the content of the address in R0 to Add
MOV @ R1 & A	→	Copy the content of the address in R1 to A register
MOV @ R0, 80	→	Copy the content 80 the port o pins to the address into R0

Example of Indirect Addressing mode

It is widely used is accessing data elements of from program memory an indirect move from the location whose address in the sum of a base register (DPTR or PC) and index register accumulator. These mode facilities lookup accesses.

This instruction used for the purpose is "MOV A,@A + DPTR "

MOV DPTR, # 4200 Copy the number of 4200 to DPTR

MOV A, # 25 Copy the number 25 to A

MOV A, @ A + DPTR copy the content of 4225 to A

Note:

4200 + 25 =>4225

Another type of indexed addressing is used is the "case jump" instruction. In this case, the destination address of a jump instruction is completed as the sum of the base pointer and the accumulator data.

4.3 INSTRUCTION SET OF 8051 MICROCONTROLLER

All members of the 8051 family execute the same instructions set. The 8051 instructions set is optimized for 8-bit content application. The Intel 8051 has excellent and most powerful instructions set offers possibilities in control area, serial Input/Output, arithmetic, byte and bit manipulation.

It has 111 instructions they are

- 49 single byte instructions
- 45 two bytes instructions
- 17 three bytes instructions

The instructions set is divided into four groups, they are

- Data transfer instructions
- Arithmetic instructions

- Logical instructions
- Call and Jump instructions

4.4 DATA TRANSFER INSTRUCTIONS

There instructions are used to copy a data between different memory and register locations, #signifies immediate addressing 8051 instructions set.

Data transfer instruction

MOV <dest-byte>, <src-byte>

Function: Move byte variable

Description: The byte variable indicated by the second operand is copied into the location specified by the first operand. The source byte is not affected. No, other register or flag is affected.

This is by bar the most flexible operation fifteen combinations of source and destination addressing modes are allowed.

Example: Internal RAM locations 30H holds 40H. The value of RAM location 40H is 10H. The data present at I/P port 1 is 1100 1010B (0CAH)

```
MOV R0, #30H        ; R0 < = 30H
MOV A, @R0          ; A <= 40H
MOV R1, A           ; R1 < = 40H
MOV B, @R1          ; B <=10H
MOV @R1, P1         ; RAM (40H) <= 0CAH
MOV P2, P1          ; P2 #0CAH
```

Leaves the value 30H in register 0, 40H is both the accumulator and register 1, 10H in register B and CAH(11001010B) both in RAM location 40H and output on port2.

MOV A, Rn

Operation: (A) ←(Rn)

Example: MOV A, R2;

This instruction copies the content of the register R2 of selected register bank to the accumulator.

Before Execution		**After Execution**	
A	R2	A	R2
05	B4	B4	B4

One byte instruction

One machine cycle

MOV A, direct

Operation: MOV (A) ←(direct)

Example: MOV A, 20H; this instruction copies the content of memory location whose address is 20H to the accumulator.

Before Execution		After Execution	
A	M	A	M
0A	20	20	20

Two bytes instruction

One machine cycle

MOV A, ACC is not a valid instruction.

MOV A, @Ri

Operation: MOV (A) ← ((Ri))

Example: MOV A, @R0. This instruction copies the content of memory location whose address is specified in the register R0 from selected register bank.

One byte instruction

One machine cycle

MOV A, #data

Operation: MOV A ← #data

Example: MOV A, #25H. This instruction copies data given with in instruction (25H) into the accumulator.

Before Execution	After Execution
A	A
B2	25

Two bytes instruction

One machine cycle

4.5 EXTERNAL DATA MOVES

It is possible to expand RAM and ROM memory space by adding external memory chips to the 8051 microcontroller.

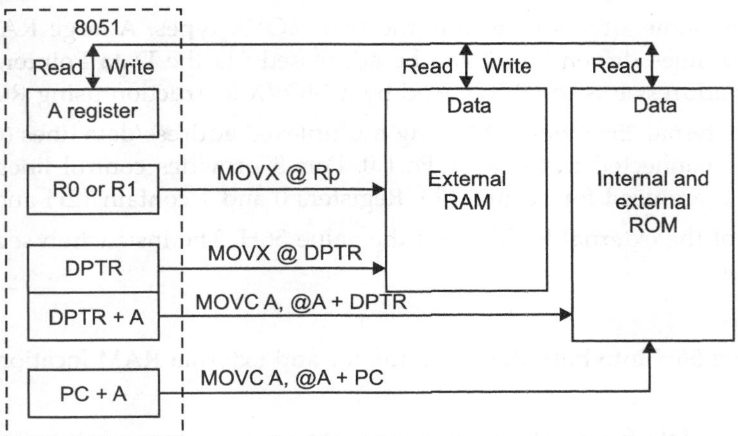

Fig. 4.1 8051 External Memory.

The external memory can be as large as 64k for each of the RAM and ROM memory areas. Opcodes that access this external memory always use indirect addressing to specify the external memory. Figure 4.1 shows the register R0, R1 and apply named DPTR can be used to hold the address of the data byte in external RAM.

R0 and R1 are limited to external RAM address range of 00H to 0FFH, while the DPTR register can address the maximum RAM space of 0000H to 0FFFFH. An X is added to the Mov mnemonics to serve as a reminder that the data move is external to the 8051.

Note :

- All external data moves must involve the A register.
- Rp can address 256 bytes; DPTR can address 64K bytes.
- MovX is normally used with external RAM or I/O addresses.
- Note that there are two sets of RAM addresses between 00H and 0FFH: one internal and one external to the 8051.

MOVX <dest-byte>, <src-byte>

Function: Move External

Description: The MOVX instructions transfer data between the Accumulator and a byte of external data memory, hence the "X" appended to MOV. There are two types of instructions, differing in whether they provide an eight-bit or sixteen-bit indirect address to the external data RAM.

In the first type, the contents of R0 or R1 in the current register bank provide an eight-bit address multiplexed with data on P0. Eight bits are sufficient for external I/O expansion decoding or for a relatively small RAM array. For somewhat larger arrays, any output port pins can be used to output higher-order address bits. These pins would be controlled by an output instruction preceding the MOVX.

In the second type of MOVX instruction, the Data Pointer generates a sixteen-bit address. P2 outputs the high-order eight address bits (the contents of DPH) while P0 multiplexes the low order eight bits (DPL) with data. The P2 Special Function Register retains its previous contents while the P2 output buffers are emitting the contents of DPH. This form is faster and more efficient when accessing very large data arrays (up to 64K bytes), since no additional instructions are needed to set up the output ports.

It is possible in some situations to mix the two MOVX types. A large RAM array with its High-order address lines driven by P2 can be addressed via the Data Pointer, or with code to output high-order address bits to P2 followed by a MOVX instruction using R0 or R1.

Example: An external 256 byte RAM using multiplexed address/data lines (e.g., an Intel 8155 RAM/I/O/Timer) is connected to the 8051 Port 0. Port 3 provides control lines for the external RAM. Ports 1 and 2 are used for normal I/O. Registers 0 and 1 contain 12H and 34H.

Location 34H of the external RAM holds the value 56H. The instruction sequence,

MOVX A, @ R1

MOVX @ R0, A

copies the value 56H into both the Accumulator and external RAM location 12H.

MOVX A, @Ri

Function : Move data from External memory into accumulator

Description : The MOVX instructions transfer data from specified external data memory location to accumulator, There are two types of instructions, differing in whether they provide an eight-bit or sixteen-bit indirect address to point the external data memory. In this case, the pointer R0 or R1 provides eight bit address.

Operation : (A) ← ((Ri))

One byte instruction.

Two machine cycle.

MOVX A, @ DPTR

Function : Move data from External memory into accumulator

Description : The MOVX instructions transfer data from specified external data memory location to accumulator, There are two types of instructions, differing in whether they provide an eight-bit or sixteen-bit indirect address to point the external data memory. In this case DPTR provides 16 bit address.

Operation: (A) ← ((DPTR))

One byte instruction.

Two machine cycle.

MOVX @Ri, A

Function: Move data from accumulator into External memory.

Description: The MOVX instructions transfer data from specified external data memory location to accumulator, there are two types of instructions, differing in whether they provide an eight-bit or sixteen-bit indirect address to point the external data memory. In this case the pointer R0 or R1 provides eight bit address.

Operation: ((Ri)) ← (A).

One byte instruction.

Two machine cycle.

MOVX @DPTR, A

Function: Move data from accumulator into External memory.

Description: The MOVX instructions transfer data from specified external data memory location to accumulator, there are two types of instructions, differing in whether they provide an eight-bit or sixteen-bit indirect address to point the external data memory. In this case DPTR provides 16 bit address.

Operation: ((DPTR)) ← (A).

One byte instruction.

Two machine cycle.

4.6 CODE MEMORY READ-ONLY DATA MOVES

Data moves between RAM locations and 8051 registers are usually of a temporary and disappears when the system is powered down.

There are times when access to a preprogrammed mass of data is needed, such as when suing tables of predefined bytes. This data must be permanent to be of repeated use and is stored in the program ROM using assemblers directives that store programmed data anywhere in ROM that the programmer wishes.

Access to this data is made possible by using indirect addressing.

MOV C A, @A +16 bit reg

The letter C is added to the Mov mnemonic to highlight the use of the opcodes for moving data.

The MOVC instructions load the Accumulator with a code byte, or constant from program memory. The address of the byte fetched is the sum of the original unsigned eight-bit Accumulator contents and the contents of a sixteen-bit base register, which may be either the Data Pointer or the PC. In the latter case, the PC is incremented to the address of the following instruction before being added with the Accumulator; otherwise the base register is not altered.

Sixteen-bit addition is performed so a carry-out from the low-order eight bits may propagate through higher-order bits. No flags are affected.

Note :

- All data is moved from the code memory accumulator.
- MOV C is normally used with internal or external ROM and can address 64K of internal or 64K of external code.

MOVC A, @A + DPTR

This instruction loads the accumulator from the contents of program memory whose address is given by the sum of the contents of accumulator and contents of DPTR register.

Operation : (A) ← (A) + (DPTR)

One byte instruction

Two machine cycle

MOVC A, @A +PC

This instruction loads the accumulator from the contents of program memory whose address is given by the sum of the contents of accumulator and contents of program counter. The current contents of program counter are incremented by 1 before summation.

Operation: (PC) ← (PC) + 1

(A) ← ((A) + (PC))

One byte instruction.

Two machine cycle.

4.7 PUSH AND POP OPCODES

Push and pop instructions are used to more the data b/w internal RAM shell and internal RAM. The operant on of the stack pointer as data is pushed or poped to the stack are in internal RAM. The stack pointer holds an internal RAM address that is called the top of the stack.

Note:

- The stack pointer is set to w 07H when the 8051 is reset, which is the same direct address in internal RAM as register R1 in bank. The first PUSH opcode would unit data to R0 of bank!. The stack pointer should be unutilized by programmer to point to an internal RAM address above the register address likely to be used by the program.

- When the stack pointer reaches FFH it "rolls over "tom 00H (R0).
- RAM ends at address 7FH: PUSH above the register bank.
- The stack pointer is usually set at address above the register bank.
- The stack pointer may be pushed and poped to the stack.

4.7.1 PUSH direct

Function: Push into stack

Description: The stack pointer is increment by one. The content of the indirect variable is than copied into the internal RAM location addressed by the stack pointer. Otherwise, no flags are affected.

Example:

On entering an interrupt routine the stack pointer 09H. The data pointer holds the value 0123H. The instruction sequence,

PUSH DPL

PUSH DPH

Will leave the stack pointer set to OBH & store 23H and 01H in internal RAM location 0AH & 0BH, respectively

Operation: PUSH

(SP) ← (SP)+1

((SP)) ← (direct)

Two bytes instruction.

Two machine cycle.

4.7.2 POP direct

Function: POP from stack

Description: The content of the internal RAM location addressed by the stack pointer is read, and the stack pointer is decrement by one. The value read is than transferred to the directly addressed bytes indirected. No, flags are affected.

Example:

The stack pointer originally contains the 32H & internal RAM location 30H through 32H contains the value 20H, 23H & 01H respectively. The instruction sequence,

POP DPH

POP DPL

Will leave the stack pointer set to 20H. Note that in this special case the stack pointer was decrement to 2FH before being loaded with the value popped (20H).

Operation : POP

(direct) ← ((stack pointer)).

(direct) ← (stack pointer)-1.

4.8 DATA EXCHANGE

MOV, PUSH and POP instructions all invdve copying the data round in the source address to the destination address the original data in the source is not changed. Exchange instruction actually move data in two directions: From source to destination and from destination to source.

Description: XCH loads the accumulator with the content of the indirected variable, at the same time they the original accumulator constants to the indirected variable.

Note:

- All addressing modes except immediate may be used in the XCH (exchange) instruction.
- All exchanges use register A.
- When using XCHD, the upper nibble of A and the upper nibble os the addresss location in Rp do not change.

XCH A, (byte)

Function: exchange accumulator with byte

Description: XCH loads the accumulator with the content of the indirected variable at the same time using the original accumulator contents to the indirected variable. The some / destination operand can use register, direct or register indirect addressing.

Example: Ro contains the address 20H. The accumulator holds the value 3Fh (00111111B0. Internal RAM location 20H holds the value 75H (01110101B).

The instruction. XCH A, @ Ro

Will leave RAM location 20H holding the values 3FH (00111111B) and 75H (01110101B) in the accumulator.

Before execution		After execution	
A	Ro (address)	A	Ro (address)
3F	75	75	3F

XCH A, (byte)

Operation: (A) ← (Rn)

→

Example: XCH A, Ro

This instruction exchange contents of accumulator with the contents of register Ro of selected register bank.

Before execution		After execution	
A	Ro (address)	A	Ro (address)
2F	72	72	2F

One byte instruction.

One machine cycle.

XCH A, direct

Operation: (A) ← (diect)

→

Example: XCH A, 40H;

This instruction exchange content of accumulator with the content of memory whose address is given with in the location (40H)

Before execution		After execution	
A	Memory (40H)	A	Memory (40H)
42	3A	3A	42

Two bytes instruction.

One machine cycle.

XCH A, @ Ri

Operation: (A) ← (Ri)

→

Example: MOV A, @Ri

This instruction exchange content of accumulator with the content of memory whose address is given by the content by register R1 of selected bank.

Before execution		After execution	
A	@R1	A	@R1
2A	4E	4E	2A

One byte instruction.

One machine cycle.

XCHD A, @ Ri

Function: Exchange digit

Description: XCHD exchange the low order nibble of the accumulator (bit 3-0) generally representing a hexadecimal or BCD digit, with that of the internal RAM location indirect addressed by the specified register. The high order nibbles (bit7-4) of each register are not affected no flages are affected.

Example: Ro contains the address 20H. The accumulator holds the value 36H (00110110B). Internal RAM location 20H holds the value 75H is (01110101B).

The instruction,

XCHD A,@Ro

Will leave RAM location 20H holding the values 75H (01110110B) and 35 (00110101B) in the accumulator.

Operation: (A3-0) → (Ri 3-0)

←

One byte instruction.

One machine cycle.

Note:

The opcode that moves data between location with in 8051 and between the 8051 and external memory have been discussed of this instruction are as follows.

Instruction Type	**Result**
MOV destination, source	Copy data from the internal RAM source

	Address to the internal RAM destination Address
MOV CA, Source	Copy internal or external program memory byte from the source is register A
MOV X destination, source	Copy byte to or from external RAM to register A
PUSH Source	Copy byte from to internal RAM stack From internal RAM source.
POP destination	Copy byte from internal RAM stack to Internal RAM destination.
XCH A, Source	Exchange data b/w register A and the Internal RAM source
XCHD A, Source	Exchange lower nibble b/w register A And the internal RAM source.

There are from addressing modes in immediate number, a register name, a direct internal RAM address, and an indirect address contained in a register.

4.9 SUMMARY

Moving Data

- WHICH MODE OF THE 8051 → immediate ,register, direct and indirect.
- All numbers must STRAT with a decimal number (0–9) or the assembler assumes the number is a label.
- Register to register moves using the register addressing mode occur between register a & R0–R7.
- R0 & R7 are limited to entered RAM address range of 00H to 0FFH.
- Only register R0 or R1 may be used for indirect addressing.
- All data is moved from the code memory to the register A.
- A push opcode copies data from the source address to the stack.
- A pop opcode copies data from the stack to the destination address.
- The SP is usually set at addresses above the register bank the SP may be pushed & poped to the stack.
- All exchanges use register A.
- Move is normally used with internal or external ROM & can address 4K of internal or 64K of entered code.
- MOV X is normally used with external RAM or I/O addresses.
- The number in register R_p must be a RAM address.

4.10 QUESTIONS

1. Data is stored at a ________address and moved to a _______address.
2. Give four addressing modes.
3. Register A, DPTR, and R0–R7 may be named as part of the________.
4. ________ and________ hold the BANK select bits, RSO and RS1.
5. Indirect addressing for MOV opcodes uses register R0 or R1 often called a ________.
6. The DPTR register can address the maximam RAM space of 0000H to 0FFFFH.
7. Data moves between RAM location and 8051 register are made by using ________ and ________opcodes.
8. Define stack pointer.
9. The first push opcode would write data to R0 of ________.
10. A ________ copy data from stack to destination address.
11. The rotate and swamp operation are limited to ________.

Chapter 5

LOGICAL OPERATIONS

5.1 INTRODUCTION

The logical instructions that perform Boolean operations.

AND
OR
EX-OR
NOT

On bytes performs the operation on a bit by bit basis. That is, if the accumulator contains 00110101B and <byte> contains 01010011B, then

ANL A, <byte>

Will leave the accumulator holding 00010001B

The rotate instructions,

RL rotate a byte to the left.

RLC rotate a byte and carry bit left.

RR rotate a byte to the right.

RRC rotate a byte and the carry to the right.

SWAP exchange the low and high nibbles in a byte.

Shift the accumulator 1-bit to the left or right. For a left rotation, the MSB rolls into the LSB position. For a right rotation, the MSB rolls into the LSB position. For a right rotation, the LSB rolls into the MSB position.

The SWAP A instruction interchanges the high and low nibbles with in the accumulator. This is a useful operation in BCD manipulations.

5.2 BYTE LEVEL LOGICAL OPERATIONS

That all such operations are done using each individual bit of the destination and source bytes. These operations, called byte-level Boolean operations. All addressing modes may be used in byte level logical operations.

List of the 8051 logical operations

ANL A, #n

ANL A, add

ANL A, Rr

ANL A, @Rp

ANL add, A

ANL add, #n

ORL A, #n

ORL A, add

ORL A, RL

ORL A, @Rp

ORL add, A

ORL add, #n

XRL A, #n

XRL A, add

XRL A, RL

XRL A, @Rp

XRL add, A

XRL add, #n

CLR A

CPL A

ANL <dest-byte>, <src-byte>

Function : Logical – AND

Description : ANL performs the bitwise logical – AND operation between the variables indicated and stores the result in the destination variable – No flags are affected.

The two operands allow six addressing mode combinations. When the destination is the accumulator, the source can use register, direct, register-indent, or immediate addressing; when the destination is a direct address, the source can be the accumulator or immediate data.

Note: When this instruction is used to modify an output port, the value used as the original port data will be read from the output data latch, no the input pins.

Example : If the accumulator holds 7.2 (01110010B) and register 0 holds 85H(10000101B) I the accumulator.

ANL A,R0

A → 72 – 01110010

R0 → 85 – 10000101

10000000

Before Execution		After Execution	
A	R0	A	R0
72	85	80	85

Will leave 80H (10000000B) in the accumulator. When the destination is a directly addresses byte, this instruction with clear combinations of bits in any RAM location or hardware register.

The instruction,

ANL P_1, #0111011B

	D_7	D_6	D_5	D_4	D_3	D_2	D_1	D_0
P_1 =	0	1	1	1	0	0	1	1

With clear bits 7, 3 and 2 of output Port 1.

½ /3 byte instruction.

1 machine cycle.

ANL A, Rn

Operation : (A) ← (A) ^ (Rn)

Example : ANL A, R2 ; Logically ANDs A and R2 and stores result in A,

A = 42 → 01000010

R2 = 79 → 01111001

01000000

Before Execution		After Execution	
A	R2	A	R2
42	79	40	79

1 byte instruction.

1 machine cycle.

ANL A, direct

Operation : (A) ← (A) ^ (direct)

Example : ANL A, 20H; Logically ANDs contents of A and memory location whose address is 20H and stores result in A.

A = A8 → 10101000

(Memory Location) 20H = 87 → 10000111

10000000

Before Execution		After Execution	
A	Memory Location 20H	A	Memory Location 20H
A8	87	80	87

2 bytes instruction.

1 machine cycle.

ANL A, @R1

Operation : (A) ← (A) ^ ((Ri))

Example : ANL A, @R1 ; Logically ANDs contents of A and memory location whose address is given by R1 and stores result in A.

Address of R0 = 55

Accumulator = 00110010

Content of RAM address at 55 = 00100010

00100010

Before Execution		After Execution	
A	Memory Location 55H	A	Memory Location 55H
32	22	22	22

1 byte instruction.

1 machine cycle.

ANL A, #data

Operation : (A) ← (A) ^ # data

Example : Logically ANDs contents of A and immediate data and stores result in A.

A = 89 → 10001001

Data = 45 → 01000101

00000001

Before Execution		After Execution	
A	Data	A	data
89	45	01	45

2 bytes instruction.

1 machine cycle.

ANL direct, A

Operation : (direct) ← (direct) ^ (A)

Example : ANL 20H, A; Logically ANDs contents of A with the contents of memory location

20H and stores result in memory location 20H.

A = 75 → 01110101

(Memory Location) 20H = 89 → 10001001

00000001

Before Execution		After Execution	
A	Memory Location 20H	A	Memory Location 20H
75	89	75	01

2 bytes instruction.

1 machine cycle.

ANL direct, #data

Operation : (direct) ← (direct) ^ #data

Example : ANL 40H, #40H; logically ANDs the contents of memory location 40H with data 40H and stores result in memory location 40H.

(Memory Location) 40H = 45 → 01000101

Data = 24 → 00100100

00000100

Before Execution		After Execution	
Memory Location 40H	Data	Memory Location 40H	data
45	24	04	24

3 bytes instruction.

2 machine cycle.

ORL<dest-byte> <src-byte>

Function : Logical – OR for byte variables

Description : ORL performs the bitwise logical – OR operation between indicated variables, storing the results in the destination byte. No flags are affected.

The two operands allow six addressing mode combinations. When the destination is the accumulator the same can user register, direct, register – indirect or immediate addressing, when the destination is a direct address, the source can be the accumulator or immediate data.

Note : When this instruction is used to modify an output port, the value used as the original port data will be read from the output data latch, not the input pins.

Example : If the accumulator holds A3H(10100011B) and R0 holds 55H (01010101B), then the instruction

ORL A, R0

A – 10100011
R0 – 01010101

11110111

Before Execution		After Execution	
A	R0	A	R0
A3	55	F7	55

Will leave the accumulator holding the value F7H.

When the destination is a directly addressed byte, the instruction can set combination of bits in any RAM locations or hardware register. The pattern of bits to be set is determined by a mask byte, which may be either a constant data value in the instruction or a variable computed in the accumulator at run-time. The instruction,

ORL P1, #00110010B

	D_7	D_6	D_5	D_4	D_3	D_2	D_1	D_0
P_1 =	0	0	1	1	0	0	1	0

Will set bits 5, 4 and 1 of output port 1.

½/3 byte instruction.

½ machine cycle.

ORL A, Rn

Operation : (A) ← (A) V (Rn)

Example : ORL A, R2; Logically ORs the contents of A and R2 and stores result in A

A = 42 → 01000010
R2 = 72 → 01110010

01110010

Before Execution		After Execution	
A	R2	A	R2
42	72	72	72

1 byte instruction.

1 machine cycle.

ORL A, direct

Operation: (A) ← (A) v (direct)

Example: ORL A, 20H; Logically ORs the contents of A and memory location 20H and stores result in A.

A = 30 → 00110000

(Memory Location) 20H = 61 → 01100000

01110000

Before Execution		After Execution	
A	Memory Location 20H	A	Memory Location 20H
30	61	70	61

2 bytes instruction.

1 machine cycle.

ORL A, @R:

Operation : (A) ← (A) v ((Ri))

Example : ORL A, @R1; Logically ORs the contents of A and memory location whose address is given by register R1 and stores result in A.

A = 35 → 00110101

(Memory Location) R1 → 40H

R1 = 22 → 00100010

00110111

Before Execution		After Execution	
A	Memory Location of R1 40H	A	Memory Location of R1 40H
35	22	37	22

1 byte instruction.

1 machine cycle.

ORL A, #data

Operation : (A) ← (A) v #data

Example: ORL A, #32H; Logically ORs the contents of A with 32H and stores result in A

A = 07 → 00000111

data = 32 → 00110010

00110111

Before Execution		After Execution	
A	Data	A	Data
07	32	37	32

2 bytes instruction.

1 machine cycle.

ORL direct, A

Operation : (direct) ← (direct) v (A)

Example : ORL, 20H, A; Logically ORs contents of A with the contents of memory location 20H and stores result of memory location 20H.

A = B2 → 10110010

(Memory Location) 20H = 07 → 00000111

10110111

Before Execution		After Execution	
A	Memory Location 20H	A	Memory Location 20H
B2	07	B2	B7

2 bytes instruction.

1 machine cycle.

ORL direct, data

Operation: (direct) ← (direct) v # data

Example: ORL 20H, #25H; Logically ORs the contents of memory location 20H and data 25H and stores result at memory location 20H.

(Memory Location) 20H = 30 → 00110000

Data = 25 → 00100101

00110101

Before Execution		After Execution	
Memory Location 20H	Data	Memory Location 20H	data
30	25	35	25

3 bytes instruction.

2 machine cycle.

XRL <dest-byte>,<src-byte>

Function: Logical exclusion-OR for byte variables.

Description: XRL performs the bitwise logical Exclusive-OR operation between the indicated variable, storing the results in the destination. No flags are affected. The two operands allow six addressing, mode combinations. When the designation is the accumulator, the source can use register, direct, register-indirect or immediate address, when the designation is a direct address, the same can be the accumulator to immediate data.

Note: When this instruction is used to modify an output port, the value used as the original port data will be read from the output data latch, not the input pins.

Example : If the accumulator holds C3H(11000011B) and R0 holds AAH(10101010B), then instruction.

XRL A, R0

A = C3 → 11000011
R0 = AA → 10101010

01101001

Before Execution		After Execution	
A	R0	A	R0
C3	AA	69	AA

Will leave the accumulator holding the value 69H(01101001B). When the designation is directly addressed byte, the instruction can complement combination of bits in any RAM location or hardware register. The pattern of bits to be complemented is then determined by a mask byte, either a constant contain in the instruction or a variable computed in the accumulator at run-time. The instruction,

XRL P1, #00110001B

Will complement bits 5, 4 and 0 of output Port1

	D_7	D_6	D_5	D_4	D_3	D_2	D_1	D_0
P_1 =	0	0	1	1	0	0	0	1

½ byte instruction.

1 machine cycle.

XRL A, Rn

Operation : (A) ← (A) V (Rn)

Example : XRL A, R2; Logically XORs the contents of A and R2 and stores result in A.

A = AB → 10101011
R2 = D2 → 11010010

01111001

Before Execution		After Execution	
A	R2	A	R2
AB	D2	79	D2

1 byte instruction.

1 machine cycle.

XRL A, direct

Operation: (A) ← (A) V (direct)

Example : XRL A, 20H; Logically XORs the contents of A with memory location 20H and stores result in A.

A = 3C → 00111100
(Memory Location) 20H = 42 → 01000010

01111110

Before Execution		After Execution	
A	Memory Location 20H	A	Memory Location 20H
3C	42	7E	42

2 bytes instruction.

1 machine cycle.

XRL A, @Ri

Operation: (A) ← (A) V ((Ri))

Example : XRL A,@R2; Logically XORs the contents of A and the memory location whose address is given R1 and stores result in A.

A = 72 → 01110010
(Memory Location) R2 = 4A → 01001010

00111000

Before Execution		After Execution	
A	Memory Location R2	A	Memory Location R2
72	4A	38	4A

1 byte instruction.

1 machine cycle.

XRL A, #data

Operation: (A) ← (A) V #data

Example: XRL, #40H; Logically XORs the contents of A with data 40H and stores result in A.

A = 9C → 10011100
data = 32 → 01001000

11010100

Before Execution		After Execution	
A	Data	A	Data
9C	48	D4	48

2 bytes instruction.

1 machine cycle.

XRL direct, A

Operation: (direct) ← (direct) V A

Example: XRL 20H, A; Logically XORs the contents at 20H and the A and stores the result at 30H.

A = 4E → 01001110

20H = F2 → 11110010

30H = BC → 10111100

Before Execution			After Execution		
A	Memory Location 20H	Memory Location 30H	A	Memory Location 20H	Memory Location 30H
4E	F2	00	4E	F2	BC

2 bytes instruction.

1 machine cycle.

XRL direct, #data

Operation: (direct) ← (direct) V #data

Example : XRL 30H, #40H; Logically XORs the contents at 30H and data 40H and stores the result at 30H.

(Memory location) 30H = 25 → 00100101

Data = 40 → 01000000

01100101

Before Execution		After Execution	
Memory Location 30H	Data	Memory Location 30H	Data
25	40	65	40

2 bytes instruction.

1 machine cycle.

CLR A

Function: Clear Accumulator

Description: The accumulator is cleared (all bits set on zero). No flags are affected.

Examples: The accumulator contains 95H(10010101B). The instruction,

CLR A

Will leave the accumulator set to 00H (00000000B).

Operation (A) ← 0

1 byte instruction.

1 machine cycle.

CPL A

Function: Complement Accumulator

Description: Each bit of the accumulator is logically complement (one's complement) bits which previously contained a one are changed to a zero and vice versa. No flags are affected.

Example: The accumulator contains 55H(01010101B). The instruction,

CPL A

Before Execution	After Execution
A	A
55	AA

Will leave the accumulator set to AAH (10101010B)

Operation: (A) ← ($\bar{A}$)

Note that instruction, then can use the SFR port latches us designations are ANL, ORL and XRL.

Note:

1. If the direct address designation is one of the port SFRs, the data latched in the SFR, not the pin data is used.
2. No flags are affected unless the direct address is the PSW.
3. Only internal RAM or SFRS may be logically manipulated.

5.3 BIT LEVEL LOGICAL OPERATIONS

Certain internal RAM (bytes 20 to 2F) and SRFs(A, B, IE, IP, P0, P1, P2, P3, PSW TCON SCON) can be addressed by then byte addresses or by the address of each bit within a byte.

Bit addressing is very easy, to after a single bit of a byte, in a control register. The Assembler equates bit address to labels, which make the program more readable.

The need for a RAM area is due to the ability of individual bit operation. The bit addresses are numbered 00H to 7F in the RAM area.

5.3.1 Internal RAM Bit Addresses

Internal RAM bytes address 20 to 2F is both byte and bit addressable. RAM is very efficient to store bit information, while the internal RAM are addressed as individual bit addresses.

The table shows the relation between byte and bit addresses.

Byte Address(hex)	Bit Address(hex)
20	00–07
21	08–0F
22	10–17
23	18–1F
24	20–27
25	28–2F
26	30–37
27	38–3F
28	40–47
29	48–4F
2A	50–57
2B	58–5F
2C	60–67
2D	68–6F
2E	70–77
2F	78–7F

5.3.2 SFR Bit Addresses

Specified function reg A, B, IE, IP, P0, P1, P2, P3, PSW, TCON, SCON are addressed at the byte level or at the bit level. The bit address is by using five MSB's of the direct address of SFR, together with B LSB's.

The table shows the list of bit addressable SFR's and their correspond bit addresses

SFR	Direct Address(hex)	Bit Addresses(hex)
A	0E0	0E0–0E7
B	0F0	0F0–0F7
IE	0A8	0A8–0AF
IP	0B8	0B8–0BF
P0	80	80–87
P1	90	90–97
P2	0A0	0A0–0A7
P3	0B0	0B0–0B7

PSW	0D0	0D0-0D7
TCON	88	88-8F
SCON	98	98-9F

Note: The SFRs are also bit addressable form the bit address by using the five most significant bits of the direct address for that SFR, together with the three least significant bits the identify the bit position from position 0(LSB) to 7(MSB).

- Bit 0E3H is bit-3 of the A reg.
- The assembles can also understand "more descriptive mnemonics, such as P0.5 for bit-5 of Port0, which is more formally addressed as 85H.

5.3.3 Bit Level Boolean Operations

The bit level Boolean logical opcodes operators on any addressable RAM or SFR bit.

Bit Addressable Control Register

The following table lists the Boolean bit level operation.

Mnemonic
ANL c,b
ANL c,/b
ORL c,b
ORL c,/b
CPL c
CPL b
CLR c
CLR b
MOV c,b
MOV b,c
SET B C
SET B b

Note that no flags, other the C flag are affected, unless the flag is an addressed bit.

Bit Level Logical Operation Examples

Mnemonic	Operation
SET B 00h	Bit 0 of RAM byte 20h = 1
MOV C, 00h	C = 1
MOV 7Fh, C	Bit 7 of RAM byte 2Fh = 1
ANL C,/00h	C=0; bit 0 of RAM byte 20h = 1

ORL C, 00h	C = 1
CPL 7Fh	Bit 7 of RAM byte 2Fh = 0
CLR C	C = 0
ORL C, /7Fh	C=1; bit 7 of RAM byte 2Fh = 0

5.4 ROTATE AND SWAP OPERATIONS

The Rotate instructions (RLA, RLC A, etc) shift the accumulator 1 bit to the left or right. For a left rotation, the MSB rolls into the LSB position. For a right rotation, the LSB rolls into the MSB position.

The SWAP A instruction interchanges the high and low nibbles with in the accumulator. This is a useful operand in BCD manipulations. For e.g. if the accumulator contains as binary number which is known to be less than 450, it can be quickly converted to BCD by the following code.

```
MOV b, #10
DIV  AB
SWAP A
ADD A, B
```

Directing the number by 10 leaves the tens digit in the low nibble of the accumulator and the ones digit in the B register. The SWAP and ADD instructions move the tens digit to the high nibble of the accumulator and the one's digit to the low nibble.

Mnemonic
RL A
RLC A
RR A
RRC A
SWAP A

RLA

Function: Rotate accumulator left.

Description: The eight bits in the accumulator are rotated one bit to the left. Bit 7 is rotated into the bit0 position. No flags are affected.

Example: The accumulator holds the value 02AH(00101010B). The instruction,

RLA

Before Execution		**Rotation**									**After Execution**	
Cy	A	Cy	A								Cy	A
1	2A	1	0	0	1	0	1	0	1	0	0	54
		0	0	1	0	1	0	1	0	0		

Will leave the accumulator holding the value 54H (01010100B) with the unaffected operation.

(An+1) ← (An) n = 0-6

(A0) ← (A1)

1 byte instruction.

1 machine cycle.

RLC A

Function: Rotate accumulator left through the carry flag.

Description: The eight bits in the accumulator and the carry flags are together rotated one bit to the left. Bit 7 moves into the carry flag; the original state of the carry flag moves into the bit0 position. No other flags are affected.

Example: The accumulator holds the value 02AH(00101010B). The instruction,

RLC A

Before Execution		Rotation									After Execution	
Cy	A	Cy	A								Cy	A
1	2A	1	0	0	1	0	1	0	1	0	0	55
		0	0	1	0	1	0	1	0	1		

Leaves the accumulator holding the value 55(01010101B). No flags are affected.

Operation: (An+1) ← (An)n = 0-6

(A0) ← (C)

(C) ← (A7)

1 byte instruction.

1 machine cycle.

RRA

Function: Rotate accumulator right.

Description: The eight bits in the accumulator are rotated one bit to the right bit 0 rotated into the bit 7 position. No flags are affected.

Example: The accumulator holds the value 02AH(00101010B). The instruction,

RRA

Before Execution		Rotation									After Execution	
Cy	A	Cy	A								Cy	A
1	2A	1	0	0	1	0	1	0	1	0	0	15
		0	0	0	0	1	0	1	0	1		

Leaves the accumulator holding the value 015H (00010101B) with the carry unaffected.

Operation: (An) ← (An+1) = 0-6

(A7) ← (A0)

1 byte instruction.

1 machine cycle.

RRC A

Function: Rotate accumulator right through carry flag.

Description: The accumulator holds the value 02AH(00101010B), the carry is zero. The instruction,

RRC A

Before Execution		Rotation									After Execution	
Cy	A	Cy	A								Cy	A
1	2A	1	0	0	1	0	1	0	1	0	0	95
		0	1	0	0	1	0	1	0	1		

Leaves the accumulator holding the value 95(10010101B), the carry is zero.

Operation: (An) ← (An+1) n = 0-6

(A7) ← (C)

(C) ← (A0)

1 byte instruction.

1 machine cycle.

SWAP A

Function: Swap nibbles with in the accumulator.

Description: Swap A interchanges the low and high order nibbles (four bit fields) of the accumulator (bit3-0 and bits7-4). The operation can also be through of as a four bit rotate instruction number flags are affected.

Example: The accumulator value(2A)(001010101). The instruction,

SWAP A

Before Execution	After Execution
A	A
2A	A2

Leaves the accumulator holding the A2H(10100010B).

Operation: SWAP

(A3-0) ← (A7-4)

1 byte instruction.

1 machine cycle.

5.5 BOOLEAN VARIABLE MANIPULATION INSTRUCTION

CLR bit

Description: The indicated bit is cleaned (reset to zero). No other flags are affected. ELR can operate a carry flag or any directly addressable bit.

Example: Port1 has previously been written with EAH(11101010B). The instruction,

CLR P1.3

Will leave the port set to E2 (1110 0010B)

1 byte instruction.

1 machine cycle.

CLR C

Operation (C) ← 0

1 byte instruction.

1 machine cycle.

CLR bit

Operation: CLR

(bit) ← 0

2 bytes instruction.

1 machine cycle.

SETB <bit>

Function: Set Bit

Description: SETB sets the indicated bit to one SETB can operate on the carry flag or any directly addressable bit. No other flags are affected.

Example: The carry is cleared. Output Port1 has been written with the value 34H (00110100B). The Instruction,

SET B C

SET B P1.0

Will leave the carry flag set to 1 and change the data output in Port1 to 35H (00110101B)

½ byte instruction.

1 machine cycle.

SET B C

Operation (C) ← 1

1 byte instruction.

1 machine cycle.

SET B bit

Operation: SET B

(bit) ← (1)

2 bytes instruction.

1 machine cycle.

CPL bit

Function: Complement bit

Description: The bit variable specified is complemented. A bit which had been a one is change to zero and vice-versa. No other flags are affected CLR can operate on the carry or any directly addressable bit.

Note: When this instruction is used to modify an input pin, the value used as the original data will be read from the output data latch, not the input pin.

Example: Port 1 has previously been written with FF(11111111B). The instruction,

UPL P1.1

Will leave the port set to FDH (11111101B)

½ byte instruction.

1 machine cycle.

CPL C

Operation: (C) ← (C)

CPL bit

Operation: (bit) ← (bit)

ANL C,<src - bit>

Function: Logical –AND for bit variables.

Description: If the Boolean value of the source bit is a logical 0 then clear the carry flag; otherwise leaves the carry flag in its current state. A slash ("/") preceding the operand in the assembly language indicates that the logical complement of the addressed bit is used as the source value, but the source bit itself is not affected. No flags are affected.

Only direct addressing is allowed for the source operand.

Example:

Set the carry flag if, and only if,

P1.0 = 1, ACC.7 = 1, and 0V = 0

MOV C, P1.0; Load carry with input pin state

ANL C, ACC.7; AND carry with ACC.Bit7

ANL C, /0V; AND with inverse of overflow flag.

2 bytes instruction.

2 machine cycle.

ANL C, bit

Operation: ANL

(C) ← (C) ^ (bit)

2 bytes instruction.

2 machine cycle.

ANL C, /bit

Operation: ANL

(C) ← (C) ^ (bit)

2 bytes instruction.

2 machine cycle.

MOV <dest – bit>, <src – bit>

Function: Move bit data

Description: The Boolean variables indicated by the second operand is copied into the location specified, by the first operand. One of the operands must be the carry flag; the may be any directly addressable bit. No other register flag is affected.

Example: The carry flag is originally set. The data present at input Port 3 is (11000101B). The data previously written to output Port 1 is C5H (11000101B)

MOV P1.3, C

MOV C, P3.3

MOV P1.2, C will leave carry cleared and change

Port 1 to 39 H(00111001B)

2 bytes instruction.

½ machine cycle.

MOV C, bit

Operation: MOV

(C) ← (bit)

2 bytes instruction.

1 machine cycle.

MOV bit, C

Operation: MOV

(bit) ← (C)

2 bytes instruction.

2 machine cycle.

5.6 SUMMARY

- The byte level logical operations use all 4 addressing modes for the source of a data byte
- Logical operations that use the port as a source, but not as a destination
- Only internal RAM or SFR's may be logically manipulated
- No flags are affected unless the direct address is PSW
- The bit level boolean logical opcodes operate on addressable RAM or SFR bit
- The swap instruction can be thought of as a rotation of nibbles in the register A
- Only the SFRs that have been identified as bit addressable may be used in bit operations
- The carry flags in RRC and RLC are affected
- Rotation and flag operations are limited to the register A
- The CJNE instruction that any RN register can be compared with an immediate value

5.7 QUESTIONS

1. Give two data levels.
2. Define byte level boolean operations.
3. Many of these byte levels operations use a ________ addresser.
4. The big level boolean logical opcodes operate on any addressable ________ or ________.
5. The A register can be rotated one bit position to the left or right with or without including the ________flag in the rotation.
6. Give an example of swap instruction.
7. Give any two rotate instructions.
8. Define SET B.
9. IP stands for ________.
10. List the bit addressable SFR and corresponding bit addresses.
11. The rotate and swap operations are limited to ________.

Chapter 6

ARITHMETIC OPERATIONS

6.1 INTRODUCTION

Application of microcontroller often involves performing mathematical calculation on data in order to alter program flow and modify program actions. The 24 arithmetic opcodes are grouped into the following types:

Mnemonics	Operation
INC destination	increment destination by one
DEC destination	decrement destination by one
ADD/ADDC destination source	add source to destination without/with carry flag
SUBB destination source	subtract with carry, source from destination
MUL AB	multiply the contents of register A & B
DIV AB	divide the contents of register A by the contents of register B
DAA	decimal adjust the A register

6.2 FLAGS

A key part of performing arithmetic operations is the ability to store in the certain results of those operations that affect the way in which the program operates.

The 8051 has several dedicated latches or flags that store the results of arithmetic operations. It has a form of arithmetic flags. The carry and auxiliary carry overflow and parity.

6.2.1 Instructions Affecting Flags

Instruction mnemonics	Flags affected
ADD	C AC OV
ADDC	C AC OV

ANL C, direct	C
CJN E	C
CLR C	C = 0
CPL C	C = C
DA A	C
DIV	C = 0 OV
MOV C, direct	C
MUL	C = 0 OV
ORL, C, direct	C
RLC	C
RRC	C
SET B C	C = 1
SUB B	C AC OV

1. Note that flags are all stored in the PSW any instruction that can modify a bit or byte in that register (MOV, SET B; XCH etc) changes the flag.
2. The Parity flag is affected by every instruction executed.
   (*i*) Parity flag will be set to 1 for add on of 1's in A.
   (*ii*) Parity flag will be set to 0 for even number of is in register A.

6.3 ARITHMETIC INSTRUCTION

6.3.1 Unsinged and signed Addition

Number may be unsigned or signed number for addition signed numbers use 7th bit as a sign it(MSB) remaining 0 to 6th bit, expresses the magnitude of the number for signed number 7th bit shows one for negative sign and shows zero for positive sign.

6.3.2 Unsigned Addition

Unsigned numbers make use of the carry flag to detect when the reset an ADD operation is a number larger than FFh. If the carry is set to 1 after an ADD, then the carry can be added to a higher order byte. So that the sum is not lost,

For e.g.

Decimal	Binary	Hexa
97	01100001	61H
183	10110111	B7H
280	1) 00011000	1) 18H

6.3.3 Signed Addition

Signed numbers may be added two ways additions of like signed number and addition of unlike signed numbers. If unlike signed numbers are added then it is not possible for the result to be larger than – 128d or +127d, and the sign of the result will always be corrected.

For e.g

ADD A, <src-bytes>

Function: ADD

Description: ADD adds the byte variables indicated to the accumulator, leaving the result in the accumulator. The carry and auxiliary carry flags are set, respectively. If there is carry out from bit-7 or bit-3, and cleared otherwise. When adding unsigned integers, the carry flag indicates an overflow occurred.

When adding signed integers, OV indicates a negative number produced as the sum of two positive operands or a positive sum from two negative operands.

Four source operand addressing modes are allowed:

Register, direct, register-direct, immediate.

byte: ½

cycle: 1

Example: The accumulator holds 89H (10001001B) and register 0 holds 95H (10010101B) .The instruction,

ADD A, R0

Will leave 1E (00011110B) in the accumulator with the AC flag cleared and both the carry flag and OV set to 1.

```
A = 2A  →  00101010
R0 = 04 →  00000100
 -----     -------------
   2E      00101110
 -----     -------------
```

Before Execution		After Execution	
A	R0	A	R0
2A	04	2E	04

ADD A, Rn

Operation: (A) ← (A) + (Rn)

Example: ADD A, R2; Adds contents of A and R2 and stores result in A.

```
 A  = 05
R2  = 71
      -----
      76
      -----
```

Before Execution		After Execution	
A	R2	A	R2
05	71	76	71

1 byte instruction

1 machine cycle

ADD A, direct

Operation: (A) ← (A) + (direct)

Example: ADD A, 20H: Adds contents of A and memory whose address is 20H and store result in A.

A = 25

(Memory Location) 20H = 43

Before Execution		After Execution	
A	Memory Location 20H	A	Memory Location 20H
25	43	68	43

2 bytes instruction

1 machine cycle

ADD A, @Ri

Operation: (A) ← (A) + ((Ri))

Example: ADD A, @R2; Adds contents of A and memory whose address is given by register R2, and stores result in A.

A = 02

Address of R2 = 2A

2C

Before Execution		After Execution	
A	Address of R2	A	Address of R2
02	2A	2C	2A

1 byte instruction

1 machine cycle

ADD A, #data

Operation: (A) ← (A) + #data

Example: ADD A, # 20H; Adds the contents of A and 20H.

A = 42H

Data = 20H

62H

Before Execution		After Execution	
A	Data	A	Data
42	20	62	20

2 bytes instruction

1 machine cycle

ADDC A, <src – byte>

Function: Add with carry

Description: ADDC simultaneously adds the byte variable indicated, the carry flag and the accumulator contents, leaving the result in the accumulator. The carry and auxiliary – carry flags are set respectively, if there is a carry-out from bit-7 or bit-3 and cleared otherwise. When adding unsigned integers, the carry flag indicates an overflow occurred.

OV is set if there is a carry-out of bit-6 but now of bit-7, or a carry-out of bit-7 but not out of bit-6;

Otherwise, OV is cleared. When adding signed integers, OV indicates a negative number produced as the sum of two positive operands or a positive sum from two negative operands.

Four source operand addressing modes are allowed: register, direct, register – indirect, or immediate.

½ byte instruction

1 machine cycle

Example: The accumulator holds 0C3H (11000011B) and register 0 holds 0AAH (10101010B) with the carry flag set. The instruction,

ADDC A, R0

Will leave 6EH (01101110B) in the accumulator with AC cleared and both the carry flag and or set to 1.

ADDC A, Rn

Operation: (A) ← (A) + (C) + (Rn)

Example: ADDC A, R2; Adds the contents of A, R2 and carry flag and stored result in A.

A = 15

CY = 1

R0 = 25

3B

Before Execution			After Execution		
A	CY	R0	A	CY	R0
15	1	25	3B	0	25

1 byte instruction

1 machine cycle

ADDC A, direct

Operation: (A) ← (A) + (C) + (direct)

Example: ADDC A, 20H; adds the contents of A, memory location whose address is 20H and the carry flag and stores result in A.

A = 14

CY = 1

(Memory Location) 20H = 25

3A

Before Execution			After Execution		
A	CY	Memory Location 20H	A	CY	Memory Location 20H
14	1	25	3A	0	25

2 bytes instruction

1 machine cycle

ADDC A, @Ri

Operation: (A) ← (A) + (C) + ((Ri))

Example: ADDC A, @R2; Adds the contents of A, memory location whose address is given by register R2 and the carry flag and stores result in the A.

A = 1A

CY = 1

(Address of R2) 20H = A4

BF

Before Execution			After Execution		
A	CY	(Address of R2) 20H	A	CY	(Address of R2) 20H
1A	1	A4	BF	0	A4

1 byte instruction

1 machine cycle

ADDC A, #data

Operation: (A) ← (A) + (C) + #data

Example: ADDC A, #20H; Adds the contents of A and carry flag and 20H and stores result in A.

A = B5
CY = 1
Data = 20

D6

Before Execution			After Execution		
A	CY	Data	A	CY	Data
B5	1	20	D6	0	20

2 bytes instruction
1 machine cycle

6.4 SUBTRACTION

SUBB A, <src-byte>

Function: Subtract with borrow.

Description: SUBB subtracts the indicated variable and the carry flag together from the accumulator, leaving the result in the accumulator, SUBB sets the carry (borrow) flag if a borrow is needed for bit-7 and clears otherwise. If C was set before executing a SUBB instruction, this indicates that a borrow was needed for the previous step in a multiple precision subtraction. So, the carry is subtracted from the accumulator along with the source operand1. AC is set if a borrow is needed for bit 3 and cleared otherwise. OV is set if a borrow is needed into bit-6, but not into bit-7 or into bit but not bit-6.

When subtracting signed integers OV indicates a negative number produced when a negative value is subtracted from a positive value or a positive result when a positive number is subtracted from a negative number.

The source operand allows four addressing modes: register, direct, register-indirect, or immediate.

Example: The accumulator holds C9H (11001001B), register, 2 holds 54H (01010100B) and the carry flag is set. The instruction,

SUBB A, R2

A = C9
R2 = 59

75

Before Execution		After Execution	
A	R2	A	R2
C9	59	75	59

2 bytes instruction

1 machine cycle

Will leave the value 74H (01110100B) in the accumulator, with the carry flag and AC cleared but OV Set.

Notice that C9H minus 54H is 75H. The difference between this and the above results is due to the carry (borrow) flag being set before the operation. If the state of the carry is not known before starting a single or multiple-precision subtraction, it should be explicitly cleared by a CLR C instruction.

SUBB A, Rn

Operation: (A) ← (A) – (C) – (Rn)

Example: SUBB A, R3; Subtracts contents of R3 and carry together from A and stores results in A.

A = 45
R3 = 25
CY = 1

19

Before Execution			After Execution		
A	R3	CY	A	R3	CY
45	25	1	19	25	1

1 byte instruction

1 machine cycle

SUBB A, direct

Operation: (A) ← (A) – (C) – (direct)

Example: SUBB A, 20H; Subtracts the contents of memory location 20H and carry together from A and stores result in A.

A = 45
R3 = 25
CY = 1

19

Before Execution			After Execution		
A	R3	CY	A	R3	CY
45	25	1	19	25	1

2 bytes instruction

1 machine cycle

SUBB A, @Ri

Operation: (A) ← (A) – (C) – ((Ri))

Example: SUBB A, @R2; subtracts the contents of memory location whose address is given by R2 and carry together from A and stores result in A.

A = 45

R3 = 25

CY = 1

19

Before Execution			After Execution		
A	R3	CY	A	R3	CY
45	25	1	19	25	1

1 byte instruction

1 machine cycle

SUBB A, #data

Operation: (A) ← (A) – (C) - #data

Example: SUBB A, #20H; Subtracts 20H from A and stores result in A.

A = 77

Data = 23

54

Before Execution		After Execution	
A	Data	A	Data
77	23	54	23

After executing SUBB A, #data the accumulator 77 and data 23 and CY flag is cleared.

2 bytes instruction

1 machine cycle

6.5 INCREMENT AND DECREMENT INSTRUCTIONS

INC<byte>

Function: Increment

Description: INC increment the indicated variable by 1. An original value of 0FFH will overflow to 00H. No flags are affected. Three addressing modes are allowed. Register, direct or register-indirect.

Note: When this instruction is used to modify an output port, the value used as the original port data will be read from the output data latch, not the input pins.

Example: Register 0 contains 7EH (011111111). Internal RAM locations 7EH and 7FH contain 0FFH and 40H respectively. The instruction sequence,

```
INC @R0
INC  R0
INC @R0
```

Will leave register0 set to 7FH and internal RAM locations 7EH and 7FH holding (respectively) 00H and 41H.

½ byte instruction.

1 machine cycle.

INC Rn

Operation: $(A) \leftarrow (A) + 1$

Rn $\rightarrow$ 5F

Before Execution	After Execution
Rn	Rn
5F	60

1 byte instruction.

1 machine cycle.

INC direct

Operation: INC

$$(direct) \leftarrow (direct) + 1$$

Add a 1 to content of the direct memory address. Assume initially the internal memory location 25H contains data 60H. After executing INC 25H, the location will have the updated value 26H.

2 bytes instruction.

1 machine cycle.

INC @Ri

Operation: $((Ri)) \leftarrow ((Ri)) + 1$

Example: INC @ R1; increment contents of memory location whose address is given by register R1 by 1.

Note Ri may be R0 or R1

Assume the pointer R0 contains 42H and the internal memory location 42 contains 2AH. After executing INC @ R0, the internal memory location 42H will have updated value 2BH.

1 byte instruction.

1 machine cycle.

INC DPTR

Function: Increment Data pointer

Description: Increment the 16-bit data pointers by 1. A 16-bit increment (modulo 2^{16}) is performed; an overflow of the low order byte of the data pointer (DPL) from FFH to 00H will increment the high-order byte DPH. No flags are affected.

Example: Registers DPH and DPL contain 12H and FEH respectively. The instruction sequence,

INC DPTR

INC DPTR

INC DPTR

Will change DPH and DPL to 13H and 01H

Operation: (DPTR) ← (DPTR) + 1

1 byte instruction.

2 machine cycles.

DEC byte

Function: Decrement

Description: The variable indicate is decremented by 1. An original value of 00H will underflow to 0FFH. No flags are affected. Four operand addressing modes are allowed: accumulator, register-direct, register-indirect.

Note: Register0 contains 7FH (01111111B). Internal RAM locations 7EH and 7FH contain 00H and 40H respectively. The instruction sequence,

DEC @R0

DEC R0

DEC @R0

Will leave register 0 set to 7EH and internal RAM locations 7EH & 7FH and 7FH set to 0FFH and 3FH.

½ byte instruction.

1 machine cycle.

DEC A

Operation: (A) ← (A)–1

Subtract a 1from accumulator

Let us assume A = 23H, after executing the instruction DEC A, there will contain 22H.

1 byte instruction.

1 machine cycle.

DEC Rn

Operation: (Rn) ← (Rn)–1

Example: DEC R3; decrements the contents of R3 by 1

Before Execution	After Execution
R3	R3
3E	3D

1 byte instruction.

1 machine cycle.

DEC direct

Operation: (direct) ← (direct) –1

Example: DEC 20H; decrements the contents of memory location whose address is 20H by 1.

2 bytes instruction.

1 machine cycle.

DEC @Rn

Operation: ((R1)) ← ((Ri)) – 1

Example: DEC @R2; decrements the contents of memory location whose address is given by register R2 by 1.

1 byte instruction.

1 machine cycle.

Note:

1. No math flags are affected.
2. All 8-bit address contents overflow FFh to 00h.
3. DPTR is 16-bit; DPTR overflows from FFFFh to 0000h.
4. The 8-bit address contents underflow from 00h to FFh.
5. There is no DEC DPTR to match the INC DPTR.

6.6 MULTIPLICATION AND DIVISION

MUL AB

Function: Multiply

Description: MUL AB the unsigned eight bit integers in the accumulators and register B. The low-order byte of the sixteen-bit product is left in the accumulator and the high-order byte in B. If the product is greater than 255(FFH) the overflow flag is set; otherwise it is cleared. The carry flag is always cleared.

Example: Originally, the accumulator holds the value 80(50H). Register B holds the value 160(0A0H). The instruction,

MUL AB

Will give the product 12,800(3200H), 80B is changed to 32H (00110010B) and the accumulator is cleared. The overflow flag is set, carry is cleared.

Operation: $(A)_{7-0} \leftarrow (A) \times (B)$

$(B)_{15-8}$

A = 58H

B = 11H

Before Execution		After Execution	
A	B	A	B
58	11H	D8	05

Examples:

MOV A, #12

MOV B, #05

MUL AB

;A = 5A; B = 00

1 byte instruction.

4 machine cycles.

DIV AB

Function: Divide

Description: DIV AB divides the unsigned eight bit integer in the accumulator by the unsigned eight-bit integer in register B. The accumulator receives the integer part of the quotient; register B receives the integer remainder. The carry and or flags will be cleared.

Exception if B had originally contained 00H, the values returned in the accumulator and B register will be undefined and the overflow flag will be set. The carry flag is cleared in any case.

Example: The accumulator contains 250(0FBH or 11111010B) and B contains 18(12H or 00010010B). The instruction,

DIV AB

Will leave B in the accumulator (0DH or 00001101B) and the value 16(10H or 00010000B) in B. Since, 250 = (13 × 18) + 16 carry and OV will both be cleared.

Operation DIV

$(A)_{15-8} \leftarrow (A) / (B)$

$(B)_{7-0}$

1 byte instruction.

4 machine cycles.

6.7 DECIMAL ARITHMETIC

DA A

Function: Decimal – adjust accumulator for addition.

Description: DA A adjusts the eight bit value in the accumulator resulting form the earlier addition of two variables (each in packed BCD format), producing two four-bit digits. Any ADD or ADDC instruction may be used to perform the addition.

If accumulator bits 3-0 are greater than nine (XXXX1010XXXX1111), or if the AC flag is one, 51x is added to the accumulator producing the proper BCD digit in the low order nibble. This internal addition would set the carry flag if a carryout of the low order four-bit field propagated through all high-order bits but it would not clear the carry flag otherwise.

If the carry flag is now set, or if the four high order bits now exceed nine (1010XXXX-111XXXX) these high order bits are incremented by six, producing the proper BCD digit in the high-order nibble. Again, this would set the carry flag if there was a carry out of the high-order bits, but would not clear the carry. The carry flag thus indicates if the sum of the original two BCD variables is greater then 100, allowing multiple precision decimal addition. OV is not affected.

All of there occur during the one instruction cycle. Essentially, this instruction performs the decimal conversion by adding 00H, 06H, 60H, or 66 to the accumulator, depending on initial accumulator and PSW conditions.

Note: DA A cannot simply convert a hexadecimal number in the accumulate to BCD notation, not does

DA A apply to decimal subtraction.

Example: The accumulator holds the value 56H (01010110B) representing the packed BCD digits of the decimal number 56. Register 3 contains the value 67(01100111B) representing the packed BCD digits of the decimal number 67. The carry flag is set. The instruction sequence,

ADDC A, R3

DA A

Will first perform a standard two's complement binary addition, resulting the value 0BEH (10111110) in the accumulator. The carry and auxiliary carry flag will be cleared.

The decimal adjust instruction will then alter the accumulator to the value 24H(00100100B), indicating the packed BCD digits of the decimal number 24, the low-order two digits of the decimal sum of 56, 67 and the carry in. The carry flag will be set by the decimal adjust instruction, indicating that a decimal overflow occurred. The true sum 56, 67 and 1 is 124.

BCD variables can be incremented or decremented by adding 01H or 99H. If the accumulator initially holds 30H (representing the digit of 30 decimal), then the instruction sequence.

ADD A, #99

DA A

Will leave the carry set and 29H in the accumulator, since, 30 + 99 = 129. The low order byte of the sum can be interpreted to mean 30 – 1 = 29.

Operation: DA

IF[[$(A_{3\text{-}0})$>9] OR [(AC)] = 1]

Then $(A_{3\text{-}0}) \leftarrow (A_{3\text{-}0}) + 6$

AND

IF[[$(A_{7\text{-}4})$ > 9] OR [(C) = 1]]

Then $(A_{7\text{-}4}) \leftarrow (A_{7\text{-}4}) + 6$

1 byte instruction.

1 machine cycle.

6.8 SUMMARY

- Arithmetic
- The c, ac and 0 be flags are arithmetic flags.
- The parity flag is affected.
- The parity is an elementary error checking method.
- All shift address contents over flow from 77h to 00h.
- There is no DEC DPTR to match the INC DPTR.
- No math flags are affected.
- Multiplication operations use register A and B as both source and destination for operation.
- Division operations use register A and B as both source and destination for operation.
- There is no comma between A and B in the division mnemonic.
- Four bits are required to represent the decimal numbers from 0 – 9 (0000 – 1001) and the numbers are often called binary coded decimal (BCD).
- Only ADD and ADDC are adjusted to BCD by DAA.

6.9 QUESTIONS

1. The instruction ADD is used to add __________ operands.
2. DAA stands for __________ .
3. In an 8 bit operand bit __________ is used for the sign bit.
4. In this valid 8051 instruction? Explain your answer "MUL A, R1".
5. A general is that if the __________ flag is set, then complement the sign.
6. Define ADD instruction with example.

Chapter 7

JUMP AND CALL OPERATIONS

7.1 INTRODUCTION

A single " JMP add" instruction, but in fact there are three SJMP, JMP, and AJMP which different in the format of the destination address JMP is a generic mnemonic which can be used of the mnemonic does not can which way the jump encoded.

In all cases the programmer specifies are destination address to the assembler in the same way: as a label or as a 16bit constant. The assembler will put the destination address into the correct format for the given instruction. If the format required by the instruction will not support the distance to the specified destination address a destination out of range message is written into the list file.

A single "CALL address "instruction, but there are two of them LCALL and ACALL which differ to the CPU. CALL is a generic mnemonic which can be used if the programmer does not came which way the address is encoded.

7.2 JUMP AND CALL INSTRUCTIONS

All these jumps specify the destination address by the relative of set method, and so are limited to a jump instance of –128 to +127 bytes from the instruction.

The jump instructions are three types

- I Unconditional
- II Conditional
- III Indexed or Absolute

The conditional jumps test against zero or compare two signed or unsigned operands. The indexed jumps index tables of displacement found in the current segment. They are used is implement case statement. All jumps are program counter-relative and all displacement are measured in bytes, relative to the first byte of case instruction (recorded in saved PC).

Note: Most of the jump opcodes add signed displacement, obtained by sign extending alpha, to the unsigned PC. The only unsigned jump displacement are in the jump indexed instructions.

7.3 SJMP

The SJMP instruction coded the destination additional as a relative offset. The instruction is 2 byte long, consisting of code and the relative offset byte. The jump distance is limited to a range of –128 to +127 bytes relative to the instruction following the SJMP.

7.4 LJMP

The LJMP instruction encoded the destination address as a 16bit constant. The instruction is 3byte long; the destination address can be any where in the 64k program memory space.

7.5 AJMP

The AJMP instruction encodes the destination address as an 11bit constant. The instruction is 2 bytes long, constants of the opcode, enrich itself constant 3 of the 11 address bytes, following by another byte containing the low 8 bit of the destination address. When the instruction is executed, this 11 byte is simply substituted for the low 11 bit in the PC. The high 5 bit stays the same. Hence, the destination has to be with the same 2k block as the instruction in following the AJMP.

7.6 RELATIVE OFFSET

The destination address for these jumps is specified to the assembler by a label or by an actual address in program memory. However, the destination address assemble to a relative offset byte which is added to the (two's complement) offset byte which is added to the PC in two's complement with arithmetic if the jump is executed. The range of the jump is therefore –128 to +127 program memory byte following the instruction.

Mnemonic	Opcode	Execution time (us)
JMP adder	Jump to addr	2
JMP @ A +DPTR	Jump to A +DPTR	2
CALL addr	Call Subroutine at addr	2
RET	Return from subroutine	2
RET1	Return from interrupt	2
NOP	No Operation	1

7.7 SHORT ABSOLUTE RANGE

Absolute range makes use of the concept of dividing memory into logical divisions called "pages". Program memory is divided into a series of pages of any binary sizes, such as 256 bytes 2k, 4k etc. Or addresses from 0000H to FFFFH.

7.8 LONG ABSOLUTE PAGE

Addresses that can access the entire program space from 0000H to FFFFH use long range addressing. It requires more byte of code by specify and are relocatable only at the being of 64K pages.

7.9 JUMPS

Jumps instruction used to respond quickly to changes in conditions. 8051 has a rich set of jump that can operate at the bit and byte levels. These jumps opcodes are one reason, the 8051 is such a powerful microcontroller.

7.9.1 Bit Jumps

Bit jumps operant according to the status of the carry flag in the PSW or the status of any bit addressable location. All bit jumps are relative to PC.

The table –shows the jump instruction that list for bit conditions

Mnemonic operation:

1. JC radd — jump relative if the carry flag is set to I
2. JNC radd — jump relative if the carry flag is reset to 0
3. JB b, radd — jump relative if addressable bit is set to I
4. JNB b, radd — jump relative if addressable bit is reset to 0
5. JBC b, radd — jump relative if addressable bit is set, and clear the addressable bit to 0

Note that numbers of flags are affected unless the bit in JBC is a flag bit in the program status word. When the bit used in a JBC instruction is a port bit, the SFR latch for that port is read, tested and altered.

JC rel

Function: Jump if carry is set.

Description: If the carry flag is set, branch to the address indicated; otherwise proceed with the next instruction. The branch destination is computed by adding the signed relative displacement in the second instruction byte to the PC, after incrementing the PC twice. No flags are affected.

Example: The carry flag is cleared. The instruction sequence

```
 JC   LABEL 1
CPL   C
 JC   LABEL 2
```

Will set the carry and cause program execution to continue at the instruction identified by the label LABEL 2.

Operation:

```
  (PC) ← (PC) +2
1F (C) = 1
THEN
  (PC) ← (PC) + rel
```

Two bytes instruction.

Two machine instruction.

JB rel

Function: Jump if bit set.

Description: If the indicated bit is me, jump to the address inducted; otherwise proceed with the next instruction. The branch destination is computed by adding the signed relative –

displacement in the third instruction byte to the PC, after incrementing the PC to the first byte of the next instruction. The bit tested is no modified. No flags are affected.

Example: The data present at input port 1 is 11001010B, the accumulator holds 56(01010110B). The instruction sequence,

JB P1.2, LABEL 1

JB ACC.2, LABEL 2

Will cause program execution to branch to the instruction at label LABEL 2.

Operation: (PC) ← (PC) + 3

1F (bit) = 1

THEN

(PC) ← (PC) + rel

Three bytes instruction.

Two machine cycle.

JNB bit, rel

Function: Jump if bit not set

Description: If the indicated bit is a zero, branch to the indicated address; otherwise proceed with the instruction. The branch destination is computed by adding the signed relative – displacement in the third instruction byte to the PC, after incrementing the PC to the first byte of the next instruction. The bit tested is not modified. No flags are not modified. No flags are affected.

Example: The data present at input port 1 is 11001010B. The accumulator holds 56H (01010110B). The instruction sequence,

JNB P!.6, LABEL1

JNB ACC.3 LABEL 2

Will cause program execution to continue at the instruction label LABEL 2.

Operation: (PC) ← (PC) + 3

1F bit = 0\

THEN

(PC) ← (PC) + rel

Three bytes instruction.

Two machine cycle.

JBC bit, rel

Function: Jump if bit set and clear bit

Description: If the indicated bit is a zero, branch to the indicated address; otherwise proceed with the instruction. The branch destination is computed by adding the signed relative – displacement in the third instruction byte to the PC, after incrementing the PC to the first byte of the next instruction. The bit tested is not modified. No flags are not modified. No flags are affected.

Note: When this instruction used to test an output pin, the value used as the original data will be read from the output data latch, hold input pin.

Example: The accumulator holds 56H (01010110B). The instruction sequence

JBC ACC.3 LABEL 1

JBC ACC.2 LABEL 2

Will cause program execution to continue at the instruction identified by the label LABEL 2, with the accumulator modified to 52H (01010010B).

Operation: (PC) ← (PC) + 3

1F (bit) = 1

THEN

(bit) ← 0

(PC) ← (PC) + rel

Three bytes instruction.

Two machine cycle.

7.9.2 Byte Jumps

Byte jumps instruction tested, bytes of data. If tested condition is true, the jump is taken. If the condition is false, the instruction after the jump is executed. All bytes jumps are relative to the PC.

CJNE < dest – byte >, < scr – byte> rel

Function: Compare and jump if not equal.

Description: CJNE compare the magnitudes of the first two operands, and branches if their values are not equal. The branch destination is computed by adding the signed relative – displacement is the last instruction byte to next instruction. The carry flag is set if the unsigned integer value of <dest- byte> is less than the unsigned integer value of < scr- byte >; otherwise, the carry is cleared. Neither operand is affected.

Example: The all contain 45h. Register 7 contains 60h. The first instruction in the sequence.

```
CJNE R7, #60h, NOT_EQ
;
;.............................. ; R7=60h
;
;
NOT_EQ : JC REQ_LOW ;IF R7&<60h
; ................................ ;IF R7>60h
```

Sets the carry flag and branches to the instruction at lable NOT_EQ. By testing the carry flag, this instruction determines whether R7 is greater or lesser than 60h. If the data being presented to port1 is also 45H, then the instruction.

WAIT: CJNE A, P!, WAIT

Clears the carry flag and continues with the next instruction in sequence, since the accumulator does equal the data read from p1. (If some other value was being input on p1, the program will loop at this point until the p1 data changes to 45h).

CJNE A, direct, rel

Operation: (PC) ← (PC)+3

IF(A) < > (direct)

THEN

(PC) ← (PC)+ relative offset

IF (A) < (direct)

THEN

(C) ← 1

ELSE (C) ← 0

3 bytes instruction.

2 machine cycle.

CJNE A, #data, rel

Operation: (PC) ← (PC)+ 3

IF (A) < > data

THEN

(PC) ← (PC) + Relative offset

IF (A) < data

THEN

(C) ← 1

ELSE

(C) ← 0

3 bytes instruction.

2 machine cycle.

CJNE Rn, #data, rel

Operation: (PC) ← (PC) + 3

IF (Rn) < > data

THEN

(PC) ← (PC) + relative offset

If (Rn) < data

THEN

(C) ← 1

ELSE

(C) ← 0

3 bytes instruction.

2 machine cycle.

CJNE @R1,# data, rel

Operation: (PC) ← (PC) + 3

IF ((Ri)) < > data

THEN

(PC) ← (PC) + relative offset

IF (Ri) < data

THEN

(C) ← 1

ELSE

(C) ← 0

3 bytes instruction.

2 machine cycle.

DJNZ < byte >, (rel_ addr)

Function: Jump if NOT ZERO

Description: DJNZ decreases the location indicated by 1, and branches to the add indicated by the second operand if the resulting value is not zero. Original values of 00h will underflow to 0FFH. No flag are affected. The branch destination would be computed by adding the signed relative-displacement value in the last instruction byte to the PC, after increasing the PC to the first byte of the following instruction. The location increased may be a register or directly addressed byte.

Note: When this instruction may be used to modify n output port, the value used as the original data port will be read from the o/p data latch, not the input pins.

Example: Internal RAM locations 40H, 50H and 60H contain the values 01H, 70H & 15H respectively. The instruction sequence,

DJZN 40h, LABEL_1

DJZN 50h, LABEL_2

DJZN 60h, LABEL_3

Will cause a jump to the instruction at label LABEL_2 with the values 00h, 6fh & 15h in the three RAM locations. The first jump was not taken, because the result was zero.

The instruction provides a simple way a executing a program loop a given number of times, or for adding a moderate time delay (from 2 to 512 machine cycle) with a single instruction. The instruction sequence,

```
            MOV R2, #08
REPEAT:     CPL P1 .6
            DJZN R2, REPEAT
```

Will REPEAT P1.6 eight times, causing four output pulses to appear at bit 6 of output port 1. Each pulse will last three machine cycle, two for DJZN and to alter the pin

DJZN Rn, rel

Operation: (PC) ← (PC) + 2

(Rn) ← (Rn) –1

IF (Rn) > 0 or (Rn) < 0

THEN

(PC) ← (PC) + rel

2/3 bytes instruction.

2 machine cycle.

DJZN direct, rel

Operation : (PC) ← (PC) +2

(direct) ← (direct) –1

IF (direct) > 0 or(Rn) < 0

THEN

(PC) ← (PC) + rel

3 bytes instruction.

2 machine cycle.

JZ rel

Function: Jump if accumulator is zero.

Description: If all bits of the accumulator are zero, branch to the address indicated; otherwise proceed with the next instruction. The branch destination is computed by adding to signed relative – displacement in the second instruction bytes to the PC, after incrementing the PC twice. The accumulator is not modified. No flags are affected.

Example: The accumulator originally contains 01H. The instruction sequence,

```
 JN    LABEL 1
DEC    A
 JZ    LABEL 2
```

Will change the accumulator to 00H and cause program executions to continue at the instruction identified by the label LABEL 2.

Operation: (PC) ← (PC) +2

IF (A) = 0

THEN

(PC) ← (PC) + rel

2 bytes instruction.

2 machine cycle.

JNZ rel

Function: Jump if accumulator is not zero.

Description: If any bit of the accumulator is a one, branch to the indicated address, otherwise proceed with the next instruction. The branch destination is computed by adding the signed relative – displacement in the second instruction bytes to the PC, after incrementing the PC twice. The accumulator is not modified. No flags are affected.

Example: The accumulator originally holds 00H. The instruction sequence,

```
JNZ    LABEL 1
INC    A
JNZ    LABEL 2
```

Will set the accumulator to 01H and continued at label LABEL 2.

Operation: (PC) ← (PC) + 2
IF (A) = ? 0
THEN
(PC) ← (PC) + rel

2 bytes instruction.

2 machine cycle.

Note that if the direct address used in a DJNZ is a port, the port SFR is decremented and tested for 0.

7.9.3 Unconditional Jumps

This type of jump occurs relative without testing the condition of any bit or byte. All jump range (relative, short, and long) is found in this stamp of jumps. There are the only jumps that can jump to any location in memory.

Example of Unconditional jumps

Mnemonic

1. JMP @ A +DPTR
2. AJMP sadd
3. LJMP ladd
4. SJMP radd
5. NOP

JMP @ A +DPTR

Function: Jump indirect

Description: Add the eight – bit unsigned contacts of the accumulator with the sixteen – bit data pointer, and load the resulting sum to the program counter. This will be the address for subsequent instruction fetcher. Sixteen – bit addition is performed. (Modulo 2^16); a carry – out from the low- order eight bits propagator through the higher order bits. Neither the accumulator nor the data pointer is allocated. No flags are affected.

Example: An even number from 0 to 6 is in the accumulator. The following sequence of the instruction will branch to one of four AJMP instructions in a JUMP table stinting at JMP _ TBL:

```
MOV DPTR, # JMP_TBL
JMP @ A + DPTR
JMP_TBL:  AJMP LABEL 0
          AJMP LABEL 1
          AJMP LABEL 2
          AJMP LABEL 3
```

If the accumulator equals 04h. When starting this sequence, execution will jump to label LABEL 2. Remember that AJUMP is a two byte instruction, so the jump instruction start at every other address.

Operation: (PC) ← A + (DPTR)

1 byte instruction.

2 machine cycle.

AJMP addr 11

Function: Absolute jump

Description: AJUMP transfer program execution to the indicated address ,which is formed at the run time by concatenating the high order five bits of the PC (after incrementing the PC twice), opcode bits 7–5 and the second byte of the instruction. The destination must therefore be within the same 2k block of the program memory as the first byte of the instruction following AJMP.

Example: The label "JMPADR" is at program memory location 0125h. The instruction

AJMP JMPADR is at location 2345h and will load the PC with 0125h.

Operation: AJUMP

$(PC) \leftarrow (PC) + 2$

$(PC_{10\text{-}0}) \leftarrow$ page address

2 bytes instruction.

2 machine cycle.x

LJMP addr 16

Function: Long JUMP

Description: LJUMP causes an unconditional branch to the indicated address by a loading the higher order and lower order bytes of the PC (respectively) with the second and third instruction bytes. The destination may therefore by anywhere in the full 64k program memory address space. No flags are affected.

Example: The label "JMPADR" is assigned to the instruction at program memory location 1234H. The instruction

LJMP JMPADR

At location 0123H will load the program counters with 1234H

Operation: LJMP

$(PC) \leftarrow addr_{15\text{ - }0}$

3 bytes instruction.

2 machine cycle.

SJMP rel

Function: Short jump

Description: Program contained unconditionally to the completed by the adding the signed displacement in the second instruction byte to the PC, after incrementing the PC twice. Therefore, the range of destination allowed is from 128 bytes preceding this instruction to 127 bytes following it.

Example: The label "RELADR" is assigned to an instruction at program memory location 0123h. The instruction

SJMP RELADR

Will assemble in to location 0100H. After the instruction is executed, the PC will contain the value 0123H.

(**Note:** Under the above condition the instruction following)

SJMP will be at 102H. Therefore, the displacement byte of the instruction will be the relative offset (0123H-0102H) +21H Put another an SJMP with a displacement of 0FEH would be the instruction infinite loop.

Operation: SJMP

(PC) ← (PC) + 2

(PC) ← (PC) + rel

2 bytes instruction.

2 machine cycle.

NOP

Function: No operation

Description: Execution continues at the following instruction other than the PC, no registers or flags are affected.

Example: It is desired to produce a low-going output pulse on bit 7 of Port 2 lasting exactly 5 cycles. A simple / CLR sequence world generate in one cycle pulse, so four additional cycles must be inserted. This may be done (assuming no interrupts are enable with the instruction sequence).

CLR P2.7

NOP

NOP

NOP

NOP

SET B P2.7

Operation: (PC) ← (PC) + 1

One byte instruction.

1 machine cycle.

7.10 CALL AND SUBROUTINES

A single "CALL addr" instruction, but there are two of them – LCALL and ACALL which differ in the format in which the subroutine address is given to the CPU. CALL is a generic mnemonic which can be used if the programme does not come which wary the address is encoded. The LCALL instruction uses the 16 bit addition format and the subroutine can be anywhere in the 64K program memory space. The ACALL instruction uses the 11 bit format and the subroutine must be in the same 2k block as the instruction following the ACALL.

In any case the programmer specifies the subroutine address to the assembler in the same way: as a label or as a 16-bit constant. The assembler will put the address into the correct format for the given instructions.

Subroutines should end with a RET instruction, which returns execution to the instruction following the CALL.

7.10.1 Calls and Return

CALLS use short and long range addressing; returns have no addressing mode specified but are always long range. The following table shows examples of CALL opcodes.

Mnemonic
ACALL sadd
LCALL ladd
RET

ACALL Address11

Function: Absolute call

Description: ACALL unconditionally calls a subroutine loaded at the indicated address. The instructions increments the PC twice to obtain the address of the following instruction, then pushes the 16-bit result onto the stack (low order byte first) and increments the stack pointer twice. The destination address is obtained by successively concatenating the five high-order bits of the incremented PC, opcode bits 7-5 and the second byte of the instruction. The subroutine called must therefore start with in the same 2k block of the program memory as the first byte of the instruction following ACALL. No flags are affected.

Example: Initially, SP equals 08H. The Label "SUBRTN" is at program memory location 0345H. After executing the instruction,

ACALL SUBRTN

At location 0125H, SP will contain 0AH, internal RAM locations 09H and 0AH instruction contain 27H and 01H respectively and the PC will contain 0345H.

Operation: ACALL

$$(PC) \leftarrow (PC) + 2$$

$$(SP) \leftarrow (SP) + 1$$

$$((SP)) \leftarrow (PC_{7-0})$$

$$(SP) \leftarrow (SP + 1)$$

$$((SP)) \leftarrow (PC_{5-8})$$

$$(PC_{10-0}) \leftarrow \text{Page address}$$

2 bytes instruction.

2 machine cycle.

LCALL address16

Function: Long CALL

Description: LCALL calls subroutines located at the indicated address. The instruction adds three to the program counter to generate the address of the next instruction and then pushes the 16-bit result onto the stack (low byte first), incrementing the stack pointer by two. The high-order and low-order bytes of the PC are then loaded, respectively, with the second and third bytes of the LCALL instruction. Program execution continues with the instruction at this address. The subroutine may therefore begin anywhere in the full 64K byte program memory address space. No flags are affected.

Example: Initially, the stack pointer equals 07H. The label "SUBRTN" is assigned to program memory location 1234H. After executing the instruction,

LCALL SUBRTN

At location 0123H, the stack pointer will contain 09H, internal RAM locations 08H and 09H will contain 26H and 0H1 and the PC will contain 1234H.

Operation: LCALL

$$(PC) \leftarrow (PC) + 3$$
$$(SP) \leftarrow (SP) + 1$$
$$((SP)) \leftarrow (PC_{7\text{-}0})$$
$$(SP) \leftarrow (SP) + 1$$
$$((SP)) \leftarrow (PC_{15\text{-}8})$$
$$(PC) \leftarrow \text{address } 15\text{-}0$$

3 bytes instruction.

2 machine cycle.

RET

Function: Return from subroutine

Description: RET pops the high and low order bytes of the PC successively from the stack, decrementing the stack pointer by the two program execution continues at the resulting address, generally the instruction immediately following an ACALL or LCALL. No flags are affected.

Example: The stack pointer originally contains the value OBH. Internal RAM locations OAH and OBH contain the values 23H and 01H respectively. The instruction

RET

Will leave the stack pointer equal to the value 09H. Program execution will continue at location 123H.

Operation: RET

$$(PC_{15\text{-}8}) \leftarrow ((SP))$$
$$(SP) \leftarrow (SP) -1$$
$$(PC_{7\text{-}0}) \leftarrow ((SP))$$
$$(SP) \leftarrow (SP) - 1$$

7.11 INTERRUPTS AND RETURNS

RET1 is used to return from an interrupt service routine. The only difference between RET and RET1 is that RET1 tells the interrupt control system that the interrupt in progress is done. If there is no interrupt in progress at the time RET1 is executed, then the RET1 is functionally identical to RET.

The following table shows the interrupt subroutine addresses

Interrupt	Address (hex) called
IE0	0003

TF0	000B
1E1	0013
TFL	001B
SERIAL	0023

RET1

Function: Return from interrupt

Description: RET1 pops the high and low order bytes of the PC successively from the stack, and restores the interrupt logic to accept additional interrupts at the same priority level as the are just processed. The stack pointer is left decremented by two. No other registers are affected; the PSW is not automatically restored to its per-interrupt status. Program execution continues at the resulting address, which is generally the instruction immediately after the point at which the interrupt request was detected. If a lower-or same-level interrupt had been pending interrupt is processed.

Example: The stack pointer originally contains the value 0BH. An interrupt was detected during the instruction ending at location 0122H. The internal RAM locations 0AH and 0BH contain the values 23H and 01H respectively. The instruction,

RET1

Will leave the stack pointer equal to 09H and return program execution to location 0123H.

Operation: RET1

$$(PC_{15\text{-}8}) \leftarrow ((SP))$$
$$(SP) \leftarrow (SP) - 1$$
$$(PC_{7\text{-}0}) \leftarrow ((SP))$$
$$(SP) \leftarrow (SP)\text{-}1$$

1 byte instruction.

2 machine cycle.

7.12 MORE DETAILS ON INTERRUPTS

7.12.1 Interrupt Structure

The 8051 core provides 5 interrupt sources: 2 external interrupts, 2 timers interrupts and the serial port interrupt. What follows is an overview of the interrupt structure for the 8051. Other MCS-51 devices have additional interrupt sources and vectors as shown in Table 1. Refer to the appropriate chapters on other devices for further information on their interrupts.

7.12.2 Interrupt Enable

Each of the interrupt sources can be individually enabled or disabled by setting or clearing a bit in the SFR named IE (Interrupt Enable). This register also contains a global disable bit, which can be cleared to disable all interrupts at once. Figure 7.1 shows the IE register for the 8051.

(MSB) (LSB)

EA	—	—	ES	ET1	EX1	ET0	EX0

Enable bit = 1 enables the interrupt.
Enable bit = 0 disables it.

Symbol	Position	Function
EA	IE.7	disables all interrupts. If EA = 0, no interrupt will be acknowledged. If EA = 1, each interrupt source is individually enabled or disabled by setting or clearing its enable bit.
—	IE.6	reserved*
—	IE.5	reserved*
ES	IE.4	Serial Port Interrupt enable bit.
ET1	IE.3	Timer 1 Overflow Interrupt enable bit.
EX1	IE.2	External Interrupt 1 enable bit.
ET0	IE.1	Timer 0 Overflow Interrupt enable bit.
EX0	IE.0	External Interrupt 0 enable bit.

*These reserved bits are used in other MCS-51 devices.

Fig, 7.1 (Interrupt Enable) Register in the 8051

7.12.3 Interrupt Priorities

Each interrupt source can also be individually programmed to one of two priority levels by setting or clearing a bit in the SFR named IP (Interrupt Priority). Figure 7.2 shows the IP register in the 8051.

A low-priority interrupt can be interrupted by a high priority interrupt, but not by another low-priority interrupt. A high-priority interrupt can't be interrupted by any other interrupt source.

If two interrupt requests of different priority levels are received simultaneously, the request of higher priority level is serviced. If interrupt requests of the same priority level are received simultaneously, an internal polling sequence determines which request is serviced. Thus, within each priority level there is a second priority structure determined by the polling sequence.

Figure 7.2 shows, for the 8051, how the IE and IP registers and the polling sequence work to determine which if any interrupt will be serviced.

(MSB) (LSB)

—	—	—	PS	PT1	PX1	PT0	PX0

Priority bit = 1 assigns high priority.
Priority bit = 0 assigns low priority.

Symbol	Position	Function
—	IP.7	reserved*
—	IP.6	reserved*
—	IP.5	reserved*
PS	IP.4	Serial Port interrupt priority bit.
PT1	IP.3	Timer 1 interrupt priority bit.
PX1	IP.2	External Interrupt 1 priority bit.
PT0	IP.1	Timer 0 interrupt priority bit.
PX0	IP.0	External Interrupt 0 priority bit.

*These reserved bits are used in other MCS-51 devices.

Fig. 7.2 Register in the 8051

7.12.4 Interrupt Control System

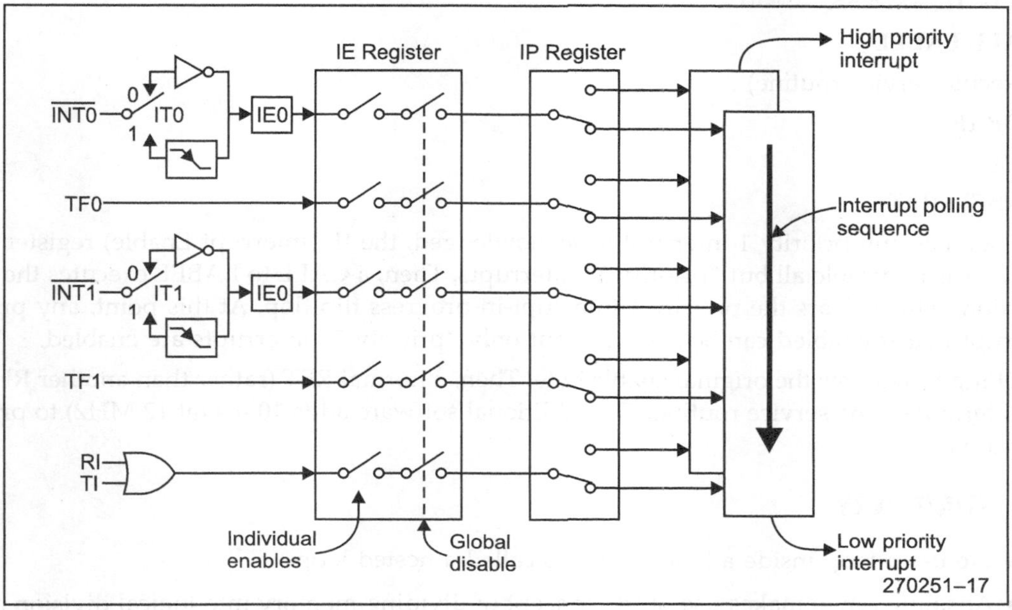

Fig. 7.3 8051 Interrupt Control System

In operation, all the interrupt flags are latched into the interrupt control system during State 5 of every machine cycle. The samples are polled during the following machine cycle. If the flag for an enabled interrupt is found to be set (1), the interrupt system generates an LCALL to the appropriate location in Program Memory; unless some other conditions block the interrupt. Several conditions can block an interrupt, among them that an interrupt of equal or higher priority level is already in progress.

The hardware-generated LCALL causes the contents of the Program Counter to be pushed onto the stack and reloads the PC with the beginning address of the service routine. As previously noted (Figure 3), the service routine for each interrupt begins at a fixed location.

Only the Program Counter is automatically pushed onto the stack, not the PSW or any other register. Having only the PC be automatically saved allows the programmer to decide how much time to spend saving which other registers. This enhances the interrupt response time, albeit at the expense of increasing the programmer's burden of responsibility. As a result, many interrupt functions that are typical in control applications -toggling a port pin, for example, or reloading a timer, or unloading a serial buffer - can often be completed in less time than it takes other architectures to commence them.

7.12.5 Simulating A Third Priority Level In Software

Some applications require more than the two priority levels that are provided by on-chip hardware in MCS-51 devices. In these cases, relatively simple software can be written to produce the same effect as a third priority level.

First, interrupts that are to have higher priority than 1 are assigned to priority 1 in the IP (Interrupt Priority) register. The service routines for priority 1 interrupts that are supposed to be interruptible by "priority 2" interrupts are written to include the following code:

```
PUSH IE
MOV IE, #MASK
CALL LABEL
(Execute service routine)
POP IE
RET
LABEL: RETI
```

As soon as any priority 1 interrupt is acknowledged, the IE (Interrupt Enable) register is re-defined so as to disable all but "priority 2" interrupts. Then, a CALL to LABEL executes the RETI instruction, which clears the priority 1 interrupt-in-progress flip-flop. At this point, any priority 1 interrupt that is enabled can be serviced, but only "priority 2" interrupts are enabled.

POPing IE restores the original enable byte. Then, a normal RET (rather than another RETI) is used to terminate the service routine. The additional software adds 10 ms (at 12 MHz) to priority 1 interrupts.

7.13 SUMMARY

- We use a loop inside a loop, which is called a nested loop.
- Absolute range makes use of the concept of dividing memory into logical division called pages.
- All jump addresses such as ADDA and ADDR.
- Must be within +127d, –127d of the instruction following the jump opcode.
- DJNZ decrement first, then checks for 0. A location set is 00H and then decremented goes to FFH then FEH and so no down on 00H.
- CJNE does not change the contents of any register or ram location. It can change the carry flag to, if the destination byte is less than the source byte.
- JMP @ a TDPTR does not change a, DPTR or any flags.
- A sub routine is a program that may be used many times in the execution of a larger program.
- Use the LCALL instruction if your subroutines are normally placed at the end of your program.
- CJNE combines a compare & a jump into one compact instruction.

7.14 QUESTIONS

1. LJMP is an ________ long jump.
2. SJMP is a______ byte instruction.
3. LJMP is a _______ byte instruction.
4. _______ is a 3 byte instruction.
5. Define nested loop?
6. What is the advantage of using SJMP over LJMP?

7. All bits, jumps are related to

   (*a*) Program counter (*b*) interrupt register (*c*) PSW's carry flage (*d*) *a* and *c*

8. In absolute jump ranges ____.

True or false:

9. All 8051 jumps are short jumps.
10. All conditional jumps are short jumps.
11. In the 8051 concept can be transferred anywhere within the 64k bytes of code space if using the LCALL instruction.
12. The target of a short jump is within –128 to +127 bytes of the current program counter.

Chapter 8

The 8255 Programmable I/O Interface

8.1 INTRODUCTION

In this chapter, we are going to study programmable peripheral Interface (PPI) 8255 designed by Intel. It is a general purpose programmable I/O device used for parallel data transfer. It has 24 I/O pins which can be grouped in three 8 bit parallel ports. Port A, port B, and port C. The eight bits of port C can be used as individual bits or be grouped in two 4bit ports: C_{upper} (C_U) and C_{lower} (C_L).

It can be programmed in two basic modes: Bit Set / Reset (BSR) mode and I/O mode. The BSR mode is used to set or reset the bits in Port C

The I/O made is further divided into three modes.

Mode 0 : single I/O

Mode 1 : I/O with hand shake

Mode 2 : bi-directional point I/O data transfer

8.2 FEATURES OF 8255 A

The 8255A is a widely used, programmable parallel I/O device.

- It can be programmed to transfer data under various conditions.

From single I/O to interrupt I/O.

- It is compatible TTL compatible.
- It has three 8-bit ports: port A, port B and port C, which are arranged in two groups of 12 pins.
- Its bits set / reset mode allows, setting and resetting of individual bits of port C.
- The 8255 can operate in three I/O modes: Mode 0, Mode 1 and Mode 2.

All I/O pins of 8255 have 2.5 mA DC driving capacity (*i.e.,* souring current of 2.5mA).

8.3 PIN DIAGRAM OF 8255 A

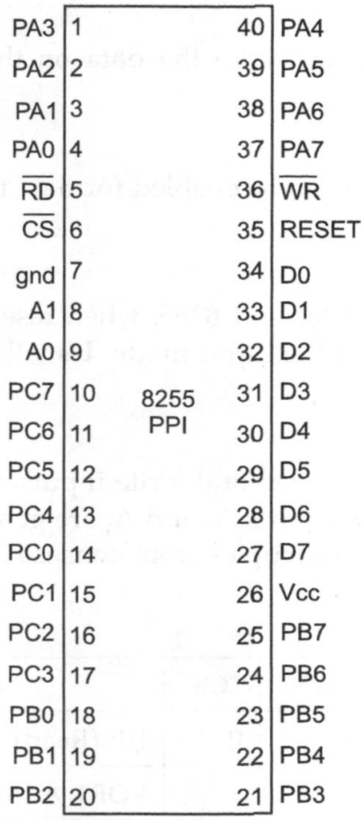

Fig. 8.1 Shows the pin diagram of 8255.

8.3.1 Explanation of Pinout Diagram

DATA BUS (PIN27-PIN34):

There bi-directional, tri state data bus lines are connected to the system data bus. They are used to transfer data and control word from microprocessor (8085) or Microcontroller (8051) to 8255 or to receive data or status word from 8255 to the 8051 and 8085.

PORT A (P_{A0} – P_A) [Pin 1-4 & Pin 37-40]:

There 8 bit bi-directional Input/Output pins are used to send data to output device and to receive data from input device. It functions as an 8bit data output latch / buffer, when used in output mode and an 8bit data input buffer, when used in input mode.

PORT B (P_{B0}-P_{B7}) (PIN_{18}-PIN_{25}):

There 8bit bi-directional input pins are used to send data to output device and to receive data from input device. It functions as 8bit data output latch/buffer when used in output mode and on 8bit data in input buffer – when used in input mode.

PORT C (P_{C0}-P_{C7}) (PIN_{10}-PIN_{17}):

There 8bit bi-directional input pins are divided into two groups P_{CL}(P_{C3}-P_{C0})and P_{C4}(P_{C7}-P_{C4}) these groups individually can transfer data in or out when transformed for simple inputs and used as handshake signals when transformed for handshake for bi-directional modes.

RD (read) (pin5):

When this pin is low, the CPU can read the data in the ports or the states word through the data buffer.

WR (write) (pin 36):

When this pin is low the CPU can write the data on the ports or in the control register through the data bus buffer.

CS (chip select) (pin 6):

This is an active low input which can be enabled for data transfer operator between the CPU and the 8255.

RESET (pin 35):

This is an active high input used to reset 8255, when reset input is high the control register is cleaned and all the ports are set to the input mode. Usually reset output signal from 8085 or 8051 is used to reset 8255.

A_0 and A_1 (Pin8, 9):

These input signal and along with read and write inputs control the selection of the control/ states word register or one of the three parts A_0 and A_1 are generally connected to the A_0 ,A_1 pins of the address bus; the 8255 therefore occupies from consecutive location in the input space

Ports and Registor Select Signals

A_1	A_0	RD	WR	CS	Operation
0	0	0	1	0	I/P (Read) Operation
					PORT A to data bus
0	1	0	1	0	I/P (Read) Operation
					PORT B to data bus
1	0	0	1	0	I/P (Read) Operation
					PORT C to data bus
0	0	1	0	0	O/P (Write) Operation
					Data bus to PORT A
0	1	1	0	0	O/P (Write) Operation
					Data bus to PORT B
1	0	1	0	0	O/P (Write) Operation
					Data bus to PORT C
1	1	1	0	0	O/P (Write) Operation
					Data bus to control register
X	x	x	x	1	Disable Function
					Data bus Tri-stated

1	1	0	1	0	Disable Function
					Illegal condition
X	x	1	1	0	Disable Function
					Data bus Tri-stated

Data Bus Buffer

This three state bi-directional 8bit buffer is used to interface the 8255 to the system data bus. Data is transmitted or received by the buffer upon execution of input or output instructions by the CPU. Control words and status informations are also transferred through the data bus buffer.

8.4 READ/WRITE AND CONTROL LOGIC

The function of this block is to manage all of the internal and external transfers of both data and Control or status words. It accepts input from the CPU Address and Control busses and in turn, issues commands to both of the Control Groups.

(CS) Chip select: A "low" on this input pin enables the communication between the 8255 and the CPU.

(RD) Read: A "low" on this input pin enables 8255 to send the data or status information to the CPU on the data bus. In essence, it allows the CPU to "read from" the 8255.

(WR) Write: A "low" on this input pin enables the CPU to write data or control words into the 8255.

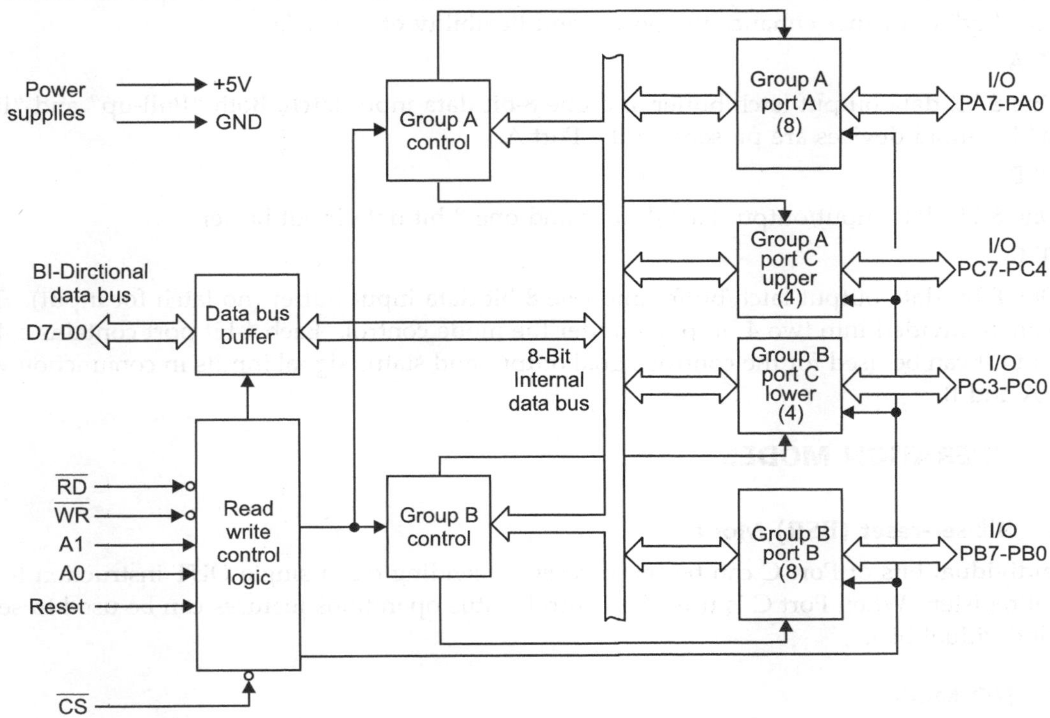

Fig. 8.2 Read/Write and Control Logic

(A_0 and A_1) Port Select 0 and Port Select 1. These input signals, in conjunction with the RD and WR inputs, control the selection of one of the three ports or the control word register. They are normally connected to the least significant bits of the address bus. (A_0 and A_1).

(RESET) Reset: A "high" on this input initializes the control register to 9Bh and all ports (A, B, C) are set to the input mode.

A_1	A_0	Selection
0	0	Port a
0	1	Port b
1	0	Port c
1	1	control

Group A And Group B Controls

The functional configuration of each port is programmed by the system software. In essence, the CPU "outputs" a control word to the 8255. The control word contains information such as "mode", "bit set", etc. That initializes the functional configuration of the 8255.Each of the Control blocks(GROUP A and GROUP B)accepts "commands" from the Read/Write control logic, receives "control words" from the internal data bus and issues the proper commands to its associated ports.

PORTS A, B and C

The 8255 contains three 8-bit ports (A, B and C)all can be configured to a wide variety of functional characteristics by the system software but each has its own special features or "personality" to further enhance the power and flexibility of the 8255.

PORT A

One 8-bit data output latch/buffer and one 8-bit data input latch. Both "Pull-up" and "Pull-down" bus-hold devices are present on the Port A.

PORT B

One 8-bit data input/output latch/buffer and one 8-bit data input buffer.

PORT C

One 8-bit data output latch/buffer and one 8-bit data input buffer (no latch for input). This port can be divided into two 4-bit ports under the mode control. Each 4-bit port contains a 4-bit latch and it can be used for the control signal output and status signal inputs in conjunction with ports A and B.

8.5 OPERATION MODES

8.5.1 Bit set-reset (BSR) Mode

The individual bits of Port C can be set or reset by sending out a single OUT instruction to the control register. When Port C is used for control/status operations features can be used to set or reset individual bits.

8.5.2 I/O Modes

Mode 0 : Simple Input/Output

In this mode, ports A and B are used as two simple 8-bit I/O ports and Port C as two 4-bit ports. Each port (or half port, in case of C) can be programmed to function as simply an input port or an output port. The input/output features in Mode 0 are as follows

1. Outputs are latched.
2. Input are buffered, not latched.
3. Ports do not have handshake or interrupt capability.

Mode 1: Input/Output with handshake

In this mode, input or output data transfer is controlled by hand shaking signals. Handshaking signals are used to transfer data between devices whose data transfer speeds are not same. For example, computer can send data to the printer with large speed but printer can't accept data and print data with this rate.

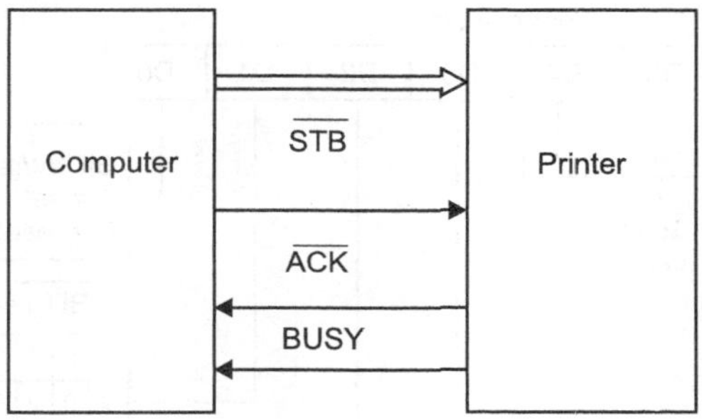

Fig. 8.3 Input/Output with Handshake

So, computer has to send data with the speed with which printer can accept. This type of data transfer is achieved by using hand shaking signals along with data signals. Fig 8.3 shows data transfer between computer and printer using handshaking signals.

These handshaking signals are used to tell computer whether printer is ready to accept the data or not. If printer is ready to accept the data, then after sending data on data bus, computer uses another handshaking signal (STB) to tell printer that valid data is available on the data bus.

The 8255 mode 1 which supports handshaking has following features.

- Two ports (A and B) functions as 8-bit I/O ports. They can be configured either as input or as output ports.
- Each port uses three lines from port C as handshake signals. The remaining two lines of port C can be used for simple I/O function.
- Input and output data are latched.
- Interrupt logic is supported.

Mode 2: Bi-directional I/O data transfer

This mode allows bi-directional data transfer (transmission and reception) over a single 8-bit data bus using handshaking signals. This feature is available only in Group A with Port A as the 8-bit bidirectional data bus and PC3 – PC7 are used for handshaking purpose. In this mode, both inputs and outputs are latched. Due to use of a simple 8-bit data bus for bidirectional data transfer, the data sent out by the CPU through Port A appears on the bus connecting it to the peripheral, only when the peripherals requests it.

The remaining lines of Port C *i.e.,* PC0 – PC2 can be used simple I/O function. This Port B can be programmed in mode 0 or in mode 1. When Port A is programmed in mode 1, PC0 - PC2 lines of Ports C are used as hand shaking signals.

8.6 CONTROL WORD FORMATS

A high on the RESET pin causes all 24 lines of the three 8-bit ports to be in the input mode. All flip flops are cleared and the interrupts are reset. This condition is maintained even after the RESET goes low. The ports of the 8255 can then be programmed for any other mode by writing a single control word into the control register, when required.

8.6.1 For Bit Set/Reset Mode

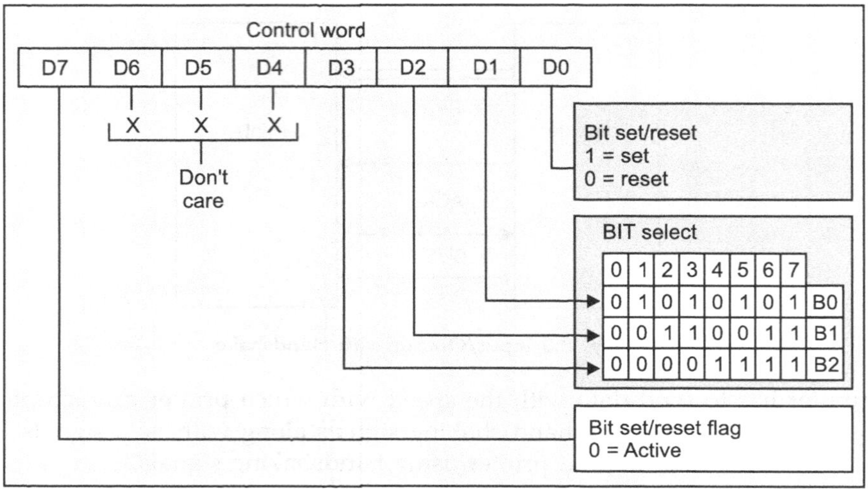

Fig. 8.4 shows bit set/reset control word format

The eight possible combinations of the status of bits D_3-D_1 (B2,B1,B0) in the Bit Set-Reset format (BSR) determine particular bit in PC0 – PC7 being set or reset as per the status of bit D_0. A BSR word is to be written for each bit that is to be set or reset. For example if bit PC3 is to be set and bit PC4 is to be reset, the appropriate BSR words that will have to be loaded into the control register will be 0xxx0111 and 0xxx1000 respectively, where x is don't care.

The BSR word can also be used for enabling or disabling interrupt signal generated by Port C when the 8255 is programmed for Mode 1 or 2 operations. This is done by setting or resetting the associated bits of the interrupts.

8.6.2 For I/O Modes

The mode definition format for input mode is shown in figure 8.5. The control words for both, mode definition and bit set-reset are loaded into the same control register, with bit D_7 used for specifying whether the word loaded into the control register is a mode definition word or Bit Set-Reset word.

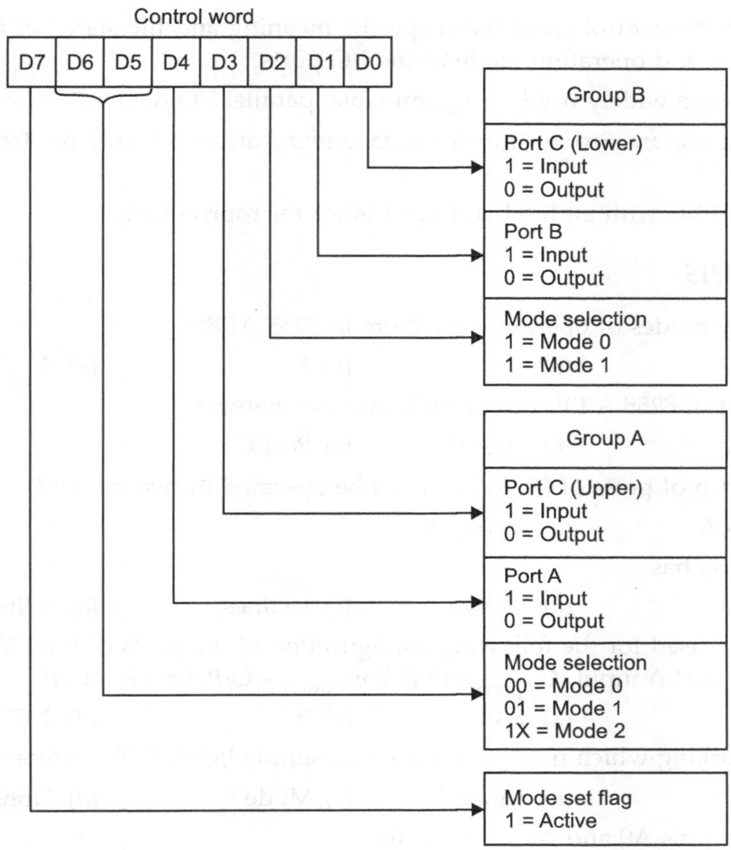

Fig. 8.5 I/O Modes

If D7 is high the word is taken as a mode definition word and if it is low. If it is taken as a Bit Set-Reset word. The appropriate bits are set or reset depending on the type of operation desired and loaded into the control register.

8.7 SUMMARY

- It has 24 I/O pins which can be grouped in three 8-bit parallel ports: Port A, Port B and Port C. The eight bits of port C can be used a individual bits or be grouped in two 4-bit ports: C_{upper} (C_u) and C_{lower} (C_L).
- It can be programmed in two basic modes: Bit Set/Reset (BSR) mode and I/O mode.
- The BSR mode is used to set or reset the bits in port C.
- The I/O mode is further divided into three modes :
- Mode 0 : Simple Input/Output.
- Mode 1 : Input/Output with handshake.
- Mode 2 : Bi-directional I/O data transfer.
- The function of I/O pins (input or output) and modes of operation of I/O ports can be programmed by writing proper control word in the control word register.

- Each bit in the control word has a specific meaning and the status of these bits decides the function and operating mode of the I/O ports.
- The 8255A is a widely used, programmable, parallel I/O device.
- It can be programmed to transfer data under various conditions, from simple I/O to interrupt I/O.
- It is compatible with all Intel and most other microprocessor.

8.8 QUESTIONS

1. How many modes of operation are there in 8255 APPI?
   (*a*) 1 (*b*) 2 (*c*) 3 (*d*) 4
2. Which part of 8255 A PPI can be split into two halves?
   (*a*) Port-A (*b*) Port-B (*c*) Port-C
3. Which group of ports of 8255 A PPI can be operated in two modes?
   (a) Group A (*b*) Group B
4. An I/O Ports has
   (*a*) 8 lines (*b*) 3 lines (*c*) 10 lines (*d*) 6 lines
5. The control used for the following configuration of the ports of Intel 8255 A for mode ø operation: Port A input, $C_{(upper)}$ – O/P, $Port_{(lower)}$ – O/P, Port B – O/P
   (*a*) Ø1 (*b*) 8 Ø (*c*) ff (*d*) 9 Ø
6. In 8255 working which mode is used when simple I/O & O/P devices
   (*a*) Mode 1 (*b*) Mode 2 (*c*) Mode 0 (*d*) None of these
7. In 8255 the pins A0 and A1 are used to
   (*a*) Select the ports and control register (*b*) Activate 8255
   (*c*) Deactivate 8255 (*d*) None of these
8. Peripherals I/O instructions are
   (*a*) Single byte (*b*) Two bytes (*c*) Three bytes (*d*) None of these
9. Which part of port C 8255 A can be grouped with port A
   (*a*) Upper port C (*b*) Lower port C
   (*c*) All lines of port C (*d*) None of these
10. Which port has no dual operation?

Chapter 9

8051 APPLICATION

9.1 INTRODUCTION

This chapter begins with interfacing application commonly used in Industrial environments. These application include such examples as the scanned LED displays, the matrix key board and memory later these examples are used as components for a system design that deals primarily with designing a single board micro computer.

We will study in detail the following typical hardware configurations and their accompanying programs.

- Key board
- Displays
  1. LCD display
  2. Seven segment display
- Traffic light controller
- A/D converter
- D/A converter

9.2 KEY BOARD

The keyboard and display devices are the two main components of microcontroller-based systems. For interfacing keyboard to the microcontroller-based system, usually push button keys are used. These push button keys when pressed, bounces a few times, closing and opening the contacts before providing a steady reading, as shown in the Fig. 9.1.

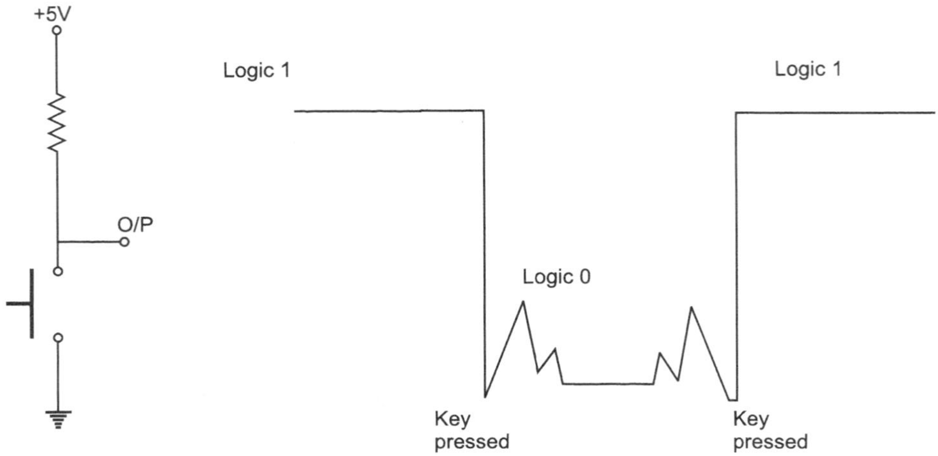

Fig. 9.1 Keyboard

9.2.1 Bouncing of Key Switch

Reading taken during bouncing period may be faulty. There fore 'Microcontroller' must wait until the key reach to a steady state; this is known as key bounce. The problem of key bounce can be eliminated using key de-bounce technique, either hardware or software.

9.2.2 Key De-bounce using hardware

Figure 9.2 Shows the circuit diagram of key bounce. It consists of 'flip-flop'.

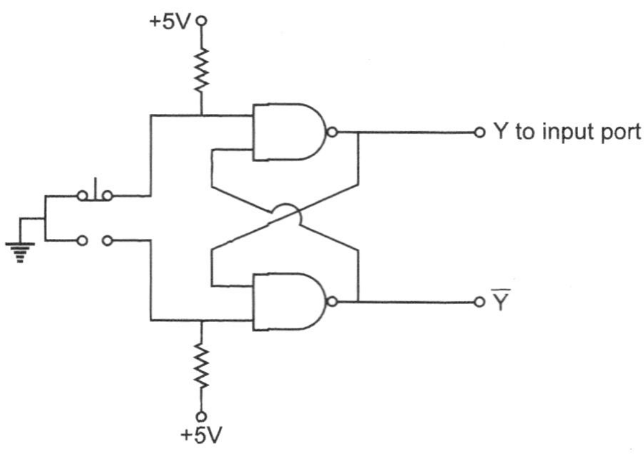

Fig. 9.2 Circuit Diagram of Keybounce

Logic is 1 when key is at position A (Unpressed)

It is logic 0 when key is at position B.

It is important note that,

When key is in between A and B, output does not change preventing of key output.

As shown in table on next page.

Key position	a	b	y
A	0	0	1
B	1	1	0
b/w A & B	1	Y	Nochange

Key position	c	d	y
A	1	1	0
B	0	0	1
b/w A & B	Y	1	Nochange

9.2.3 Key bouncing using Software

In the software technique, when a key press is found, the microprocessor waits for at least 10 ms before it accepts the key as an input. This 10ms period is sufficient to settle key at steady state. Figure 9.3 shows the flowchart with key debounce technique.

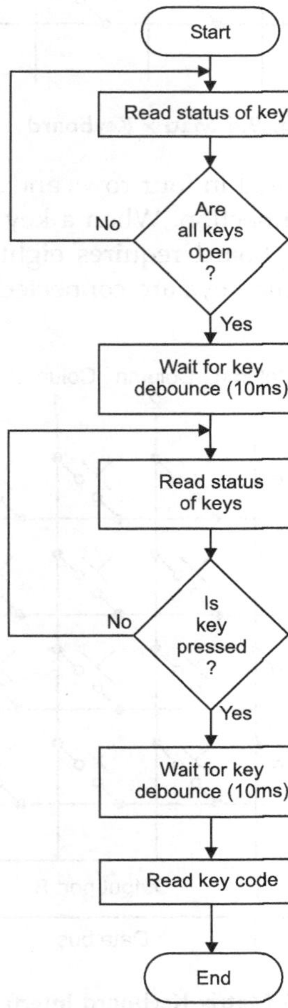

Fig. 9.3 Flow Chart with Key debounce Techniques

9.2.4 Matrix keyboard Interface

To reduce numbers of connections keys are arranged in the matrix form as shown in Fig. 9.4.

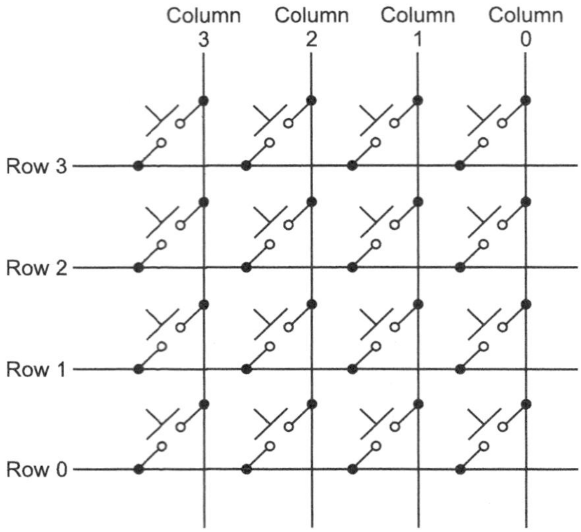

Fig. 9.4 Matrix Keyboard

Figure 9.4 shows sixteen keys arranged in four rows and four columns. When keys are open, rows and columns do not have any connection. When a key is pressed, it shorts corresponding rows and one column. This matrix keyboard requires eight lines to make all the connections instead of the sixteen lines required if the keys are connected individually.

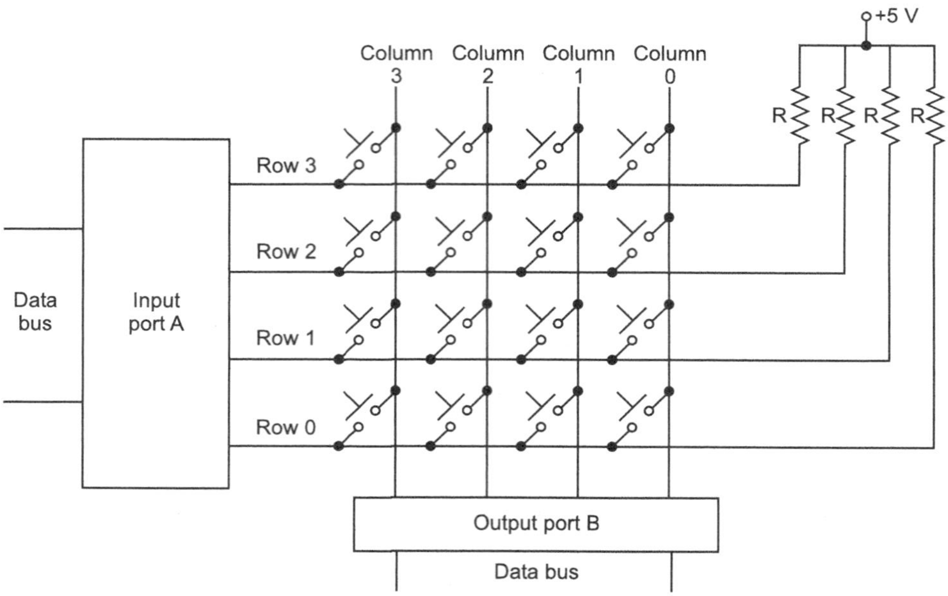

Fig. 9.5 Matrix Keyboard Interface

Figure 9.5 shows the interfacing of matrix keyboard. It requires two ports an input port and an output port. Rows are connected to the input port referred to as returned lines and columns are connected to the output port referred to as scan lines. We know that, when all keys are open, rows and columns do not have any connections. When any key is pressed it shorts corresponding row and column. If the output line of the column is low, it makes corresponding row line low; otherwise the status of row line is high. The key is identified by data sent on the output port and input code required from the input port.

1. Whether any key is pressed or not.
   (*a*) Make all column lines zero by sending low on all output lines. This activates all keys in the keyboard matrix

   Note: When scan lines are logic high, the status on the return lines does not change, it will remain logic high.

   (*b*) Read the status of the return lines. If the status of all lines is logic high, key is not pressed, otherwise key is pressed.
2. (*a*) Activate keys from any column by making any one column line zero.

   (*b*) Read the status of return lines. The zero on any return line indicates key in pressed from the corresponding row and selected columns

   If the status of all lines is logic high, key is not pressed from that column.

   (*c*) Active the keys from the next column and repeat 2 and 3 for all columns.

9.3 DISPLAY INTERFACING

Most of the microcontroller controlled instruments and machines need to display letter of the alphabet and numbers to give directions or data values to users. This information can be displayed using simple LED & LCD displays are used

9.3.1 Seven Segment Display

For the seven-segment display you can use the LT-541 or LCD-5061-11 chip. Each of the segments of the display is connected to a pin on the 8051. In order to light up a segment on the pin must be set to 0v. To turn a segment off the corresponding pin must be set to 5v. This is simply done by setting the pin on the 8051 to '1' or '0'.

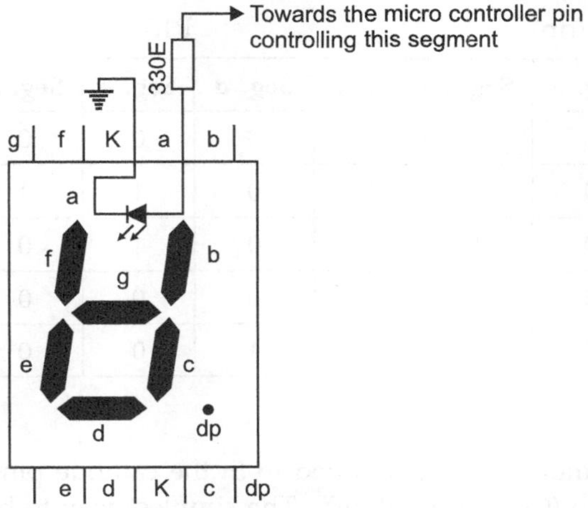

Fig. 9.6 Seven Segment Display

LED displays are

- Power-hungry (10ma per LED)
- Pin-hungry (8 pins per 7-seg display)

But they are cheaper than LCD displays

Seven segment displays are available in two types

1. Common anode
2. Common cathode

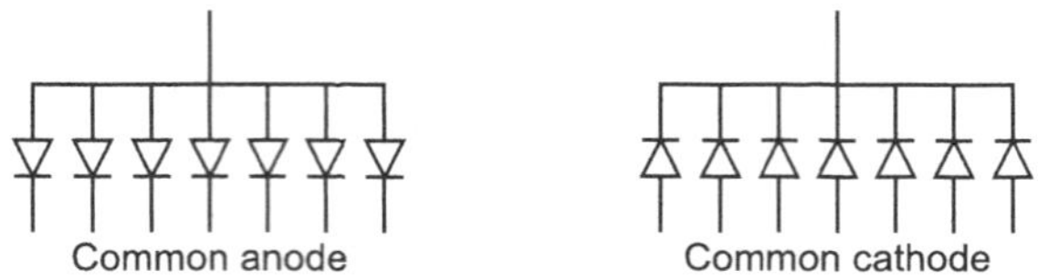

Fig. 9.7 Common Anode and Cathode

But, common anode displays are most suitable for interfacing with 8051, since 8051 port pins are sink currents better than sourcing it

Creating Digit Pattern

For display digit say 7 we need to light display to do so we have to provide logic 0(0v). At anode of these segments

CONNECTIONS

SEGMENT NUMBERS	8051 PIN NUMBER
a	P1.0
b	P1.1
c	P1.2
d	P1.3
e	P1.4
f	P1.5
g	P1.6
h(dp)	P1.7

Digit	Seg. h	Seg. g	Seg. f	Seg. e	Seg. d	Seg. c	Seg. b	Seg. a	HEX
0	1	1	0	0	0	0	0	0	C0
1	0	0	0	0	0	1	1	0	06
2	1	0	1	0	0	1	0	0	A4
3	1	0	1	1	0	0	0	0	B0
4	1	0	0	1	1	0	0	1	99

Interfacing

Note that I am using common anode pin is tied to 5v.the cathode pins are connected to port 1 through 330-ohm resistance (Current limiting). The simplest way to be drive a display is via a display driver. These are available for up to 4 displays. Alternatively displays can be driven by a microcontroller and if more than the displays are required the method of driving them is called multiplexing.

If a single display is to be driven from a microcontroller, 7 lines will be needed plus one for the decimal point. For each additional display only is extra line is needed. To produce a 4, 5, or 6 Digit display, all the seven segment displays are connected in parallel.

Each display is turned on at a rate above 100 times per second and it will appear that all the displays are turned on at the same time. As each display is turned on the appropriate information must be delivered to it so that it will give the correct reading.

For extracting the ones digit and the tens digit, macro dig byte is used. It stores the hundreds digit, the tens digit, and the ones digit into variables Dig1, Dig2, and Dig3. In our case, upon macro execution, Dig1 will equal 0, Dig2 will equal 2, and Dig3 will equal 1.

9.3.2 Interfacing To LCD Display

The most common way to accomplish this is with the LCD (liquid crystal display).LCD'S have become a cheap and easy way to get text display for an embedded systems common display are setup as 16 to 20 characters by 1 to 4 lines.

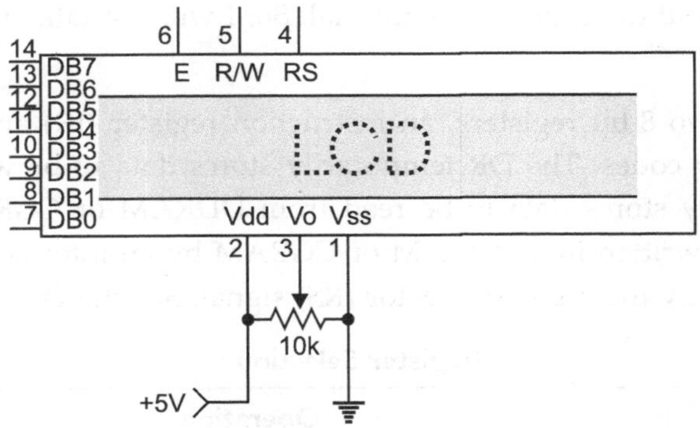

Fig. 9.8 Pin Circuit Diagram of LCD

Pinout

- **8 data pins D7:D0**

  Bi-directional data/command pins.

  Alphanumeric characters are sent in ASCII format.

- **RS: Register Select**

  RS = 0 -> Command Register is selected

  RS = 1 -> Data Register is selected

- **R/W: Read or Write**

  0 -> Write, 1 -> Read

- **E: Enable (Latch data)**

  Used to latch the data present on the data pins. A high-to-low edge is needed to latch the data.

- **VEE: contrast control**

Display Data RAM (DDRAM)

Display data RAM (DDRAM) is where you send the characters (ASCII code) you want to see on the LCD screen. It stores display data represented in 8-bit character codes. Its capacity is 80 characters (bytes). Below you see DD RAM address layout of a 2*16 LCD.

←Character position (dec.)	00	01	02	03	04	05	06	07	08	09	10	11	12	13	14	15	16	17	18	19
←Row0 DDRAM address(hex)	00	01	02	03	04	05	06	07	08	09	0A	0B	0C	0D	0E	0F	10	11	12	13
←Row1 DDRAM address(hex)	40	41	42	43	44	45	46	47	48	49	4A	4B	4C	4D	4E	4F	50	51	52	53

←Character position (dec.)	20	21	22	23	24	25	26	27	28	29	30	31	32	33	34	35	36	37	38	39
←Row0 DDRAM address(hex)	14	15	16	17	18	19	1A	1B	1C	1D	1E	1F	20	21	22	23	24	25	26	27
←Row1 DDRAM address(hex)	54	55	56	57	58	59	5A	5B	5C	5D	5E	5F	60	61	62	63	64	65	66	67

In the above memory map, the area shaded in black is the visible display (For 16x2 displays).

For first line addresses for first 15 characters is from 00h to 0Fh. But, for second line address of first character is 40h and so on up to 4Fh for the 16th character. So, if you want to display the text at specific positions of LCD, we require to manipulate address and then to set cursor position accordingly.

Character Generator RAM (CGRAM)-User defined character RAM

In the character generator RAM, we can define our own character patterns by program. CG RAM is 64 bytes, allowing for eight 5*8 pixel, character patterns to be defined. However, how to define this and use it is out of scope of this tutorial. So, I will not talk any more about CGRAM

Registers

The HD44780 has two 8-bit registers, an instruction register (IR) and a data register (DR). The IR stores instruction codes. The DR temporarily stores data to be written into DDRAM or CGRAM and temporarily stores data to be read from DDRAM or CGRAM. Data written into the DR is automatically written into DDRAM or CGRAM by an internal operation. . These two registers can be selected by the register selector (RS) signal. See the table below:

Register Selection

RS	R/W	Operation
0	0	IR write as an internal operation (display clear, etc.)
0	1	Read busy flag (DB7) and address counter (DB0 to DB6)
1	0	DR write as an internal operation (DR to DDRAM or CGRAM)
1	1	DR read as an internal operation (DDRAM or CGRAM to DR)

If a shot data bus is used, the LCD will require a total of 11 data lines.

The three control lines are EN RS & RW.

Note that the EN line must be raised/lowered before/after each instruction sent to the LCD regardless of whether that instruction is read or write text or instruction. In short, you must always manipulate EN when communicating with the LCD. EN is the LCD's way of knowing that you are talking to it. If you don't raise/lower EN, the LCD doesn't know you're talking to it on the other lines.

Checking the Busy Flag

You can use subroutine for checking busy flag or just a big (and safe) delay.

1. Set R/W Pin of the LCD HIGH (read from the LCD).
2. Select the instruction register by setting RS pin LOW.
3. Enable the LCD by Setting the enable pin HIGH.
4. The most significant bit of the LCD data bus is the state of the busy flag (1=Busy, 0=ready to accept instructions/data). The other bits hold the current value of the address counter.

If the LCD never come out from "busy" status because of some problems, the program will "hang", waiting for DB7 to go low. So, in real applications it would be wise to put some kind of time limit on the delay for example, a maximum of 100 attempts to wait for the busy signal to go low. This would guarantee that even if the LCD hardware fails, the program would not lock up.

Code Example

It is easy (and clean tech.) to make different subroutines and then call them as we need.

Busy flag checking	Data write Routine	Command write Routine
ready: SETB P1.7 ;D7 as input CLR P3.6 ;RS=0 CMD SETB P3.5 ;RW=1 for read **again:** SETB P3.7 ;H->L pulse on E CLR P3.7 JB P1.7, again ret	**data**: MOV P1, A ;move acc. data to port SETB P3.6 ;RS=1 data CLR P3.5 ;RW=0 for write SETB P3.7 ;H->L pulse on E CLR P3.7 LCALL ready ret	**command:** MOV P1, A ;move acc. data to port CLR P3.6 ;RS=0 for CMD CLR P3.5 ;RW=0 for write SETB P3.7 ;H->L pulse on E CLR P3.7 LCALL ready ret
Initialization	**Display clear**	**Displaying "HI"**

initialization:
MOV A, # 38H; Initialize, 2-lines, 5X7 matrix.
LCALL Command
MOV A, #0EH ; LCD on, cursor on
LCALL Command
MOV A, #01H ; Clear LCD Screen
LCALL Command
MOV A, #06H ; Shift cursor right
LCALL Command
	clear:
	SETB p3.7 ;enable EN
	CLR 3.6 ; RS=0 for CMD.
	MOV DATA, #01h
	CLR p3.7 ; disable EN
	LCALL ready
	RET
	
	Note- As we need to clear the LCD frequently and not the whole initialization , it is better to use this routine separately.
		LCALL initialization
		LCALL clear
		MOV A, #'H
		ACALL data
		MOV A, #'I
		LCALL data

Let's now try code for displaying text at specific positions.

I want to display "MAHESH" in message "Hi MAHESH" at the right corner of first line then I should start from 10th character.

"HI" from here (at 00) — "Mahesh" from here (at 10)

00	01	02	03	04	05	06	07	08	09	10	11	12	13	14	15	←Character position (dec.)
00	01	02	03	04	05	06	07	08	09	OA	OB	OC	OD	OE	OF	←Row0 DDRAM address(hex)
40	41	42	43	44	45	46	47	48	49	4A	4B	4C	4D	4E	4F	←Row1 DDRAM address(hex)

So, referring to table 80h+0Ah= 8Ah.

So, below is code and I don's think that you will need explanation comments.

Assembly Language

```
lcall Initialization
lcall clear
mov a, #'H
lcall data
mov a, #'I
lcall data
mov a, #8ah
lcall command
mov a, #'M
lcall data
mov a, #'A
lcall data
mov a, #'H
lcall data
mov a, #'E
lcall data
mov a, #'S
lcall data
mov a, #'H
lcall data
```

9.4 TRAFFIC LIGHT CONTROLLER

A traffic light controller is connected to an 8051 as shown in figure 9.9.

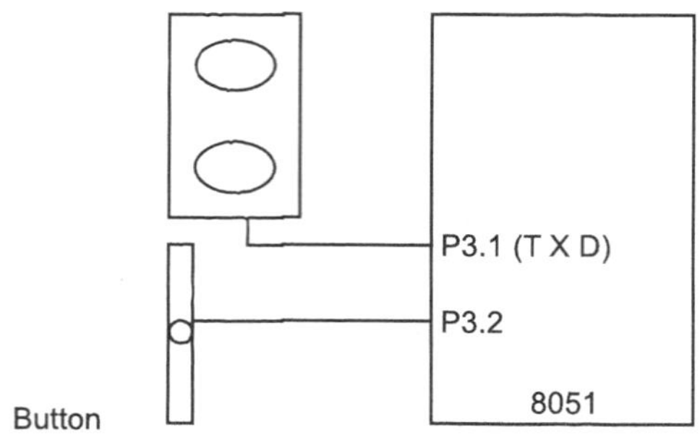

Fig. 9.9 Traffic Light Controller

The controller turns off the red light & turns on the green light for pedestrians, whenever the ASCII representation for letter "G" is transmitted serially to the controller. Similarly, it turns off the green light & turns on the red light whenever the ASCII representation for "R" is transmitted serially to the controller (4800 band rate)

Write an assembly language program to perform the following

- Whenever a pedestrian pushes the button ,wait for 5 seconds & then turn on the green light

- Keep the green light on for 20 seconds
- Then turn on the red light

If a pedestrian pushes the button during the above process no action should be taken (i.e., adjust the external interrupt triggering mechanism accordingly)

Note:

- The frequency of the crystal oscillator, xtal, is 11.0592 MHz.
- Use the timers of the 8051 for generating time delays. You can ignore the overload due to instruction in calculating the time delay. In other words, the amount of time for executing the instruction doesn't calculation.
- Use the polling techniques for serial transmission and time delay generation.

[**HINT:** Write down a subroutine called FIVESEC, to generate number of times in the interrupt service routine to generate 5 seconds & 20 seconds delays]

9.5 ADC INTERFACING

Analog signals are very common inputs to microcontroller system most transducers & sensors such as temperature, pressure, velocity humidity are analog. Therefore, we need to convert it.

Analog to digital converter:

Commonly used ADC device → ADC 0804

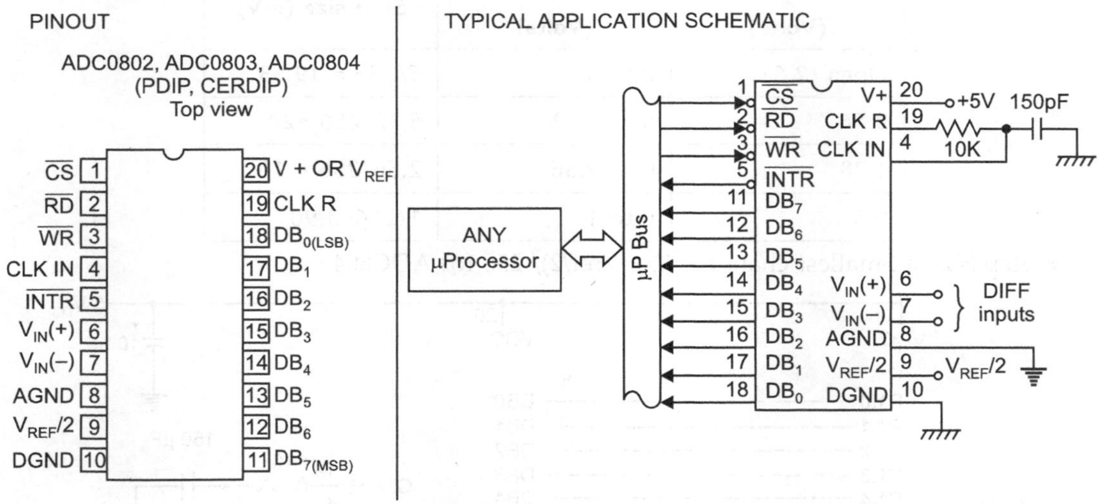

Fig. 9.10 Pinout Diagram of ADC0804

As shown in the typical circuit ADC0804 can be interfaced with microcontroller. You need a minimum of 11 pins to interface ADC0804, eight for data pins and 3 for control pins.

About Ic

PinOut

- CS – Chip Select, active low.
- RD – Read Digital data from ADC, H-L edge triggered.
- WR -- Start conversion, L-H pulse edge triggered.

- INTR -- end of conversion, Goes low to indicate conversion done.
- Data bits -- D0-D7.
- CLK IN & CLK R.

CLK IN is an input pin connected to an external clock source when an external clock is used for timing. However, ADC804 has an internal clock generator.

To use the internal clock generator of the ADC804, the CLK IN and CLK R pins are connected to a capacitor and a resistor. In that case, the clock frequency is determined by the equation.

$$f = 1/1.1RC$$

$$R = 10K \quad \text{and} \quad C = 150pF \qquad f = 606Hz$$

the conversion time is 110 μs.

Input Voltage range

- Default 0-5V. Can be changed by setting different value for Vref/2 pin.

$$Vin = Vin(+) - Vin(-).$$

- Range = 0 to 2x Vref/2.

  For Vin = 2x Vref/2. We get 256 as a digital output on D0-D7. (Refer Table)

Vref/2 (Volts)	Vin (Volts)	Step size (mV)
Open (2.5)	0 to 5	5/256 = 19.53
2.56	0 to 5.12	5.12/256 =20
1.28	0 to 2.56	2.56/256 = 10
0.5	0 to 1	1/256=3.90

- Step Size a Smallest change – (2 x Vref/2)/ 256 for ADC804

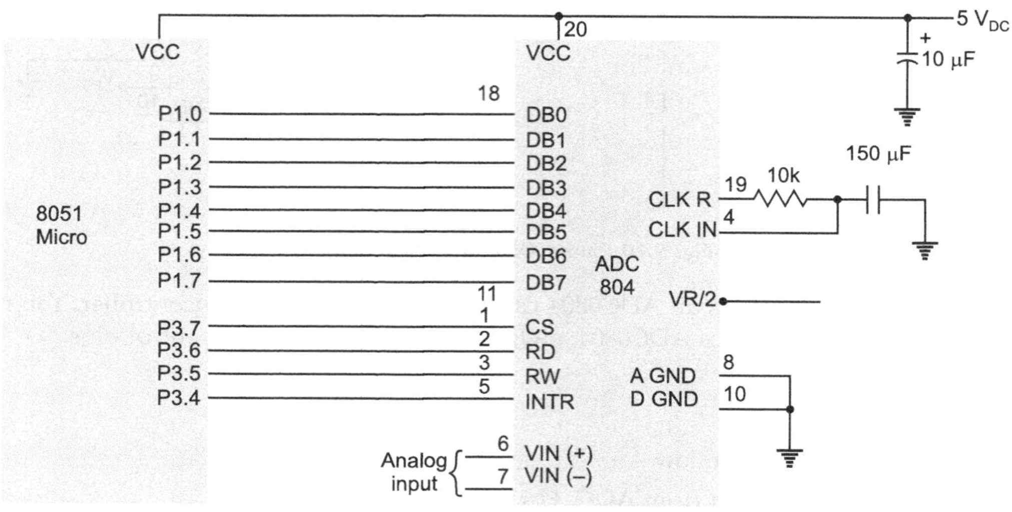

Fig. 9.11 Interfacing of ADC0804

For e.g. for step size 10mv, digital output on D0-D7 changes by one count for every 10mv change of the input analog voltage.

Data Out

Dout = Vin / Step Size

For input vtg. of 2.56 volts (Vref=1.28 volts) and stepsize of 10mv Dout =2560/10 =256 or FF that is full scale output.

Conversion Time

Greater than 110μs for ADC804.

Resolution

8 bits for ADC804.

INTERFACING ADC804 TO 8051

Signals to be interfaced (on the ADC804)

– D0-D7, RD, WR, INTR, CS

Can do both Memory mapping and IO mapping.

Memory Mapping (timing is critical)

– Connect D0-D7 of ADC804 to the data bus of the 8051 system

– Connect RD, WR of the ADC804 to the 8051 system (ensure polarity)

– Connect CS of ADC804 to an appropriate address decoder output

– Connect INTR of ADC804 to an external interrupt pin on the 8051 (INT0 or INT1)

IO Mapping (easiest - I prefer)

– Connect D0-D7, RD, WR, CS, INTR to some port bits on the 8051 (12 in all).

Algorithm

- Make CS=0 and send a low-to-high to pin WR to start the conversion.
- Keep monitoring INTR
  - If INTR =0, the conversion is finished and we can go to the next step.
  - If INTR=1, keep polling until it goes low.
- After INTR=0, we make CS=0 and send a high-to-low pulse to RD to get the data out of the ADC804 chip.

Assembly Language (A51)

```
ADC_IO:
    mov P1, #0xff ; To configure as input
AGAIN
    clr p3.7 ;Chip select
    setb P3.6 ;RD = 1
    clr P3.5 ;WR = 0
    setb P3.5 ;WR = 1 – low to high transition
WAIT:
    jb P3.4, WAIT ;wait for INTR
    clr p3.7 ;generate cs to ADC
```

```
clr P3.6 ;RD = 0 -High to low transition
mov A, P1 ;read digital o/p
sjmp AGAIN
```

Interfacing Adc804 To 8051

ADC808/809 chip with 8 analog channels. This means this kind of chip allows to monitor 8 different transducers.

- ADC804 has only ONE analog input: Vin(+).
- ALE: Latch in the address.
- Start: Start of conversion (same as WR in 804).
- OE: Output enables (same as RD in 804).
- EOC: End of conversion (same as INTR in 804).

Channel	C B A
IN0	000
IN1	001
IN2	010
IN3	011
IN4	100
IN5	101
IN6	110
IN7	111

Algorithm

Notice that the ADC808/809 that there is no self-clocking and the clock must be provided from an external source to the CLK pin. (You can use programmable clock oscillator to enable or disable clock by programmable bit.)

- Select an analog channel by provide bits to A, B, C.
- Enable clock.
- Activate ALE with a low-to-high pulse.
- Activate SC with a high-to-low pulse (start conversion) the conversion is begun on the falling edge of the start conversion pulse. You can use circuit like.
- Monitor EOC pin .After conversion this pin goes high.
- Activate OE with a high-to-low pulse to read data out of the ADC chip.

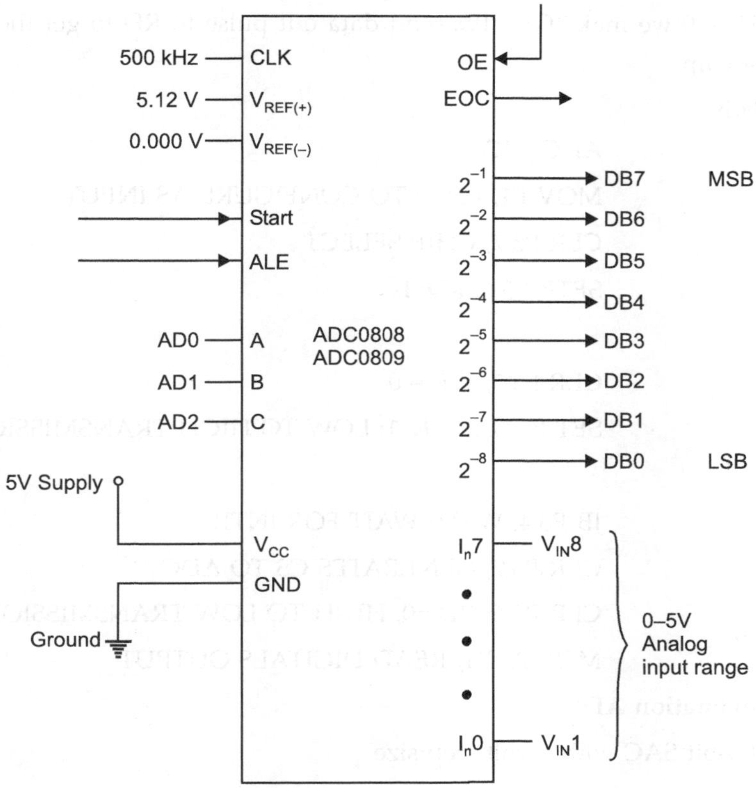

Fig. 9.12 ADC0804

Signals to be interfaced (on the ADC 0804) d0-d7, RD, WR, NTR, CS can do both memory mapping & IO mapping

Memory mapping

- Connect d0-d7 of ADC 0804 to the data bus of the 8051 system.
- Connect RD,WR of the ADC 0804is the 8051 system(ensure polarity).
- Connect CS of ADC 0804 to an appropriate address decoder output.
- Connect INTR of ADC 0804 to an external interrupt pin on the 8051(INTO or INTI).

Input Mapping

- Connect d0-d7,RD,WR,CR,INTR to some port bits on the 8051(12 in all).

Algorithm

- Make CS = 0 and send a low is high to pin WR to start the conversion.
- Keep the monitoring INTR.
- If INTR = 0, the conversion is finished and we can go to the next step.
- If INTR = 1,keep polling until it goes low.

- After INTR = 0 we make CS = 0& send data out pulse to RD to get the data out of the ADC 0804 chip.

Assembly Language

```
            ADC _ IO:
            MOV P1, #0Xff; TO CONFIGURE AS INPUT
AGAIN       CLR P3.7; CHIP SELECT
            SETB P3.6; RD=1

            CLR P3.5; WR = 0
            SET B P3.5; WR=1; LOW TO HIGH TRANSMISSION
WAIT
            JB P3.4; WAIT; WAIT FOR INTR
            CLR P3.7; GENERATES CS TO ADC
            CLR P3.6; RD =0; HIGH TO LOW TRANSMISSION
            MOV A, P1; READ DIGITALS OUTPUT
```

Successive Approximation ADC

Illustration of 4-bit SAC with 1 volt step size

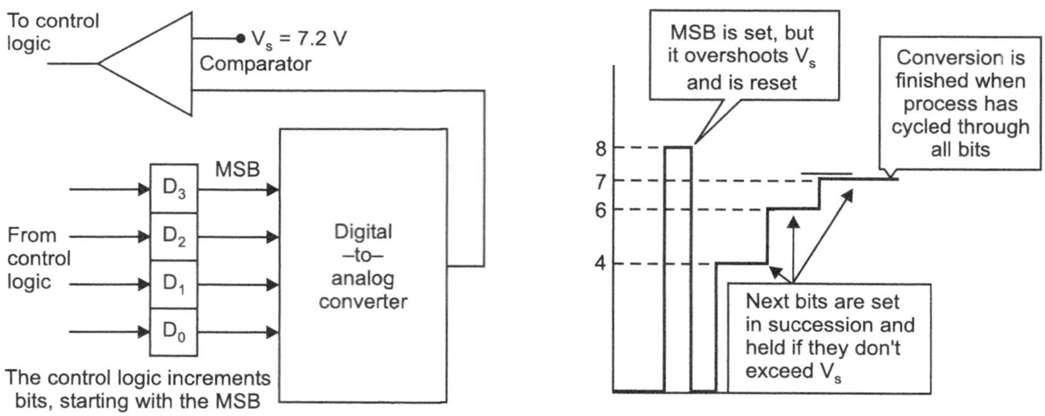

Fig. 9.13 4 bit Successive Approximation ADC

The successive approximation ADC is much faster than the digital ramp ADC, because it uses digital logic to converge on the value closest to the input voltage. A comparator and a DAC are used in the process. A flowchart emplaning the working is shown in the Fig. 9.14.

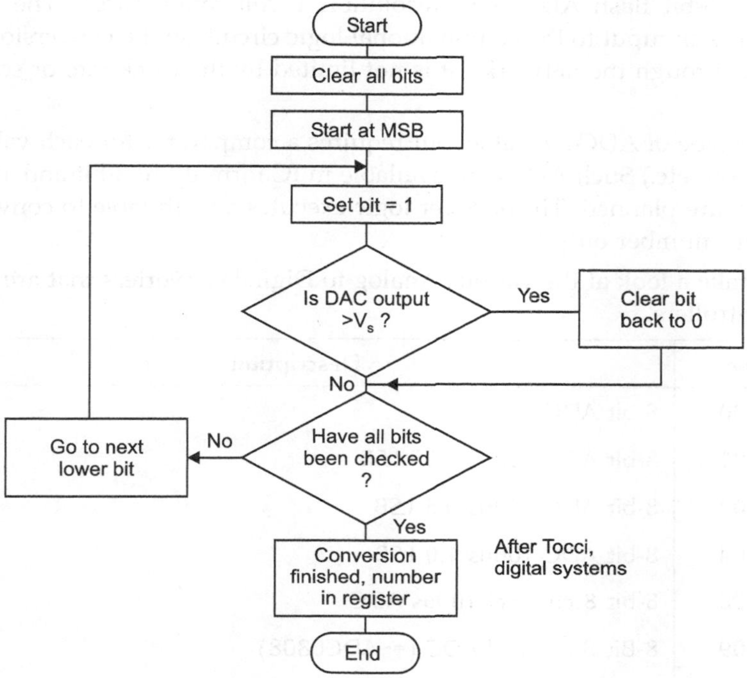

Fig. 9.14 SAC Flowchart

Flash ADC

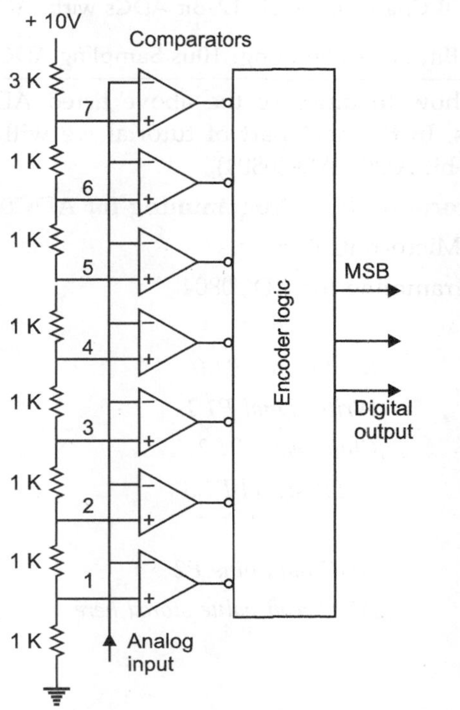

Fig. 9.15 3-bit Flash ADC

Illustrated is a 3-bit flash ADC with resolution 1 volt (after Tocci). The resistor net and comparators provide an input to the combinational logic circuit, so the conversion time is just the propagation delay through the network - it is not limited by the clock rate or some convergence sequence.

It is the fastest type of ADC available, but requires a comparator for each value of output (63 for 6-bit, 255 for 8-bit, etc.) Such ADCs are available in IC form up to 8-bit and 10-bit flash ADCs (1023 comparators) are planned. The encoder logic executes a truth table to convert the ladder of inputs to the binary number output.

Now, we lets take a look at the various Analog to Digital converters that are most commonly used with our controllers.

Name	Description
ADC0800	8-bit ADC
ADC0801	8-bit ADC 100us 0.25 LSB
ADC0802	8-bit ADC 100us 0.5 LSB
ADC0804	8-bit ADC 100us 1.0 LSB
ADC0808	8-bit 8 channel 100us ADC
ADC0809	8-Bit 8 channel ADC (=~ADC0808)
AD571	10-Bit, A/D Converter, Complete with Reference and Clock
MAX1204	5V, 8-Channel, Serial, 10-Bit ADC with 3V Digital Interface
MAX1202	5V, 8-Channel, Serial, 12-Bit ADCs with 3V Digital Interface
MAX195	16-Bit, Self-Calibrating, 10us Sampling ADC

More informations on how to interface the above listed ADC can be obtained from the datasheets of respective ICs. In the next part of tutorial we will look into the interfacing and programming of a simple 8-bit ADC (ADC0804).

ADC interfacing with Microcontrollers: Programming for ADC0804

- Programming 8051 Microcontroller
- 8051 Assembly Programming for ADC0804

CODE:

```
    rd equ P1.0             ;Read signal P1.0
    wr equ P1.1             ;Write signal P1.1
    cs equ P1.2             ;Chip Select P1.2
    intr equ P1.3           ;INTR signal P1.3

    adc_port equ P2         ;ADC data pins P2
    adc_val equ 30H         ;ADC read value stored here

    org 0H
start:                      ;Start of Program
```

```
    acall conv              ;Start ADC conversion
    acall read              ;Read converted value
    mov P3,adc_val          ;Move the value to Port 3
    sjmp start              ;Do it again

conv:                       ;Start of Conversion
    clr cs                  ;Make CS low
    clr wr                  ;Make WR Low
    nop
    setb wr                 ;Make WR High
    setb cs                 ;Make CS high
wait:
    jb intr,wait            ;Wait for INTR signal
    ret                     ;Conversion done

read:                       ;Read ADC value
    clr cs                  ;Make CS Low
    clr rd                  ;Make RD Low
    mov a,adc_port          ;Read the converted value
    mov adc_val,a           ;Store it in local variable
    setb rd                 ;Make RD High
    setb cs                 ;Make CS High
    ret                     ;Reading done
```

- Programming 8051 in C for ADC0804

CODE:

```
#include <REGX51.H>#define adc_port P2        //ADC Port
#define rd P1_0             //Read signal P1.0
#define wr P1_1             //Write signal P1.1
#define cs P1_2             //Chip Select P1.2
#define intr P1_3           //INTR signal P1.3

void conv();                //Start of conversion function
void read();                //Read ADC function

unsigned char adc_val;

void main(){
    while(1){               //Forever loop
        conv();             //Start conversion
```

```
        read();              //Read ADC
        P3 = adc_val;        //Send the read value to P3
    }
}
void conv(){
        cs = 0;              //Make CS low
        wr = 0;              //Make WR low
        wr = 1;              //Make WR high
        cs = 1;              //Make CS high
        while(intr);         //Wait for INTR to go low
}
void read(){
        cs = 0;              //Make CS low
        rd = 0;              //Make RD low
        adc_val = adc_port;  //Read ADC port
        rd = 1;              //Make RD high
        cs = 1;              //Make CS high
}
```

Note: *Keep this in mind that whenever you are working with an IC and you want to know how to communicate with that IC, and then simply look into the timing diagram of that IC from its datasheet. It gives you complete information that you need regarding the communication of IC.*

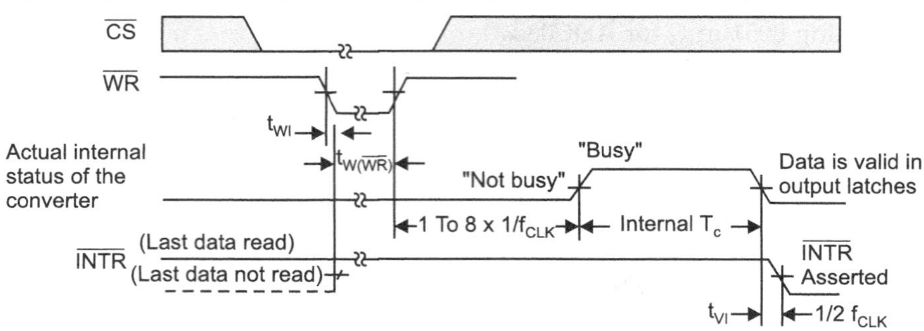

Fig. 9.16 Start Conversion

The above timing diagrams are from ADC0804 datasheet. The first diagram (Figure 9.16) shows how to start a conversion. Also, you can see which signals are to be asserted and at what time to start a conversion. So, looking into the timing diagram Figure 9.16. We note down the steps or say the order in which signals are to be asserted to start a conversion of ADC. As we have decided to make Chip select pin as low so we need not to bother about the CS signal in the timing diagram. Below steps are for starting an ADC conversion. I am also including CS signal to give you a clear picture. While programming we will not use this signal.

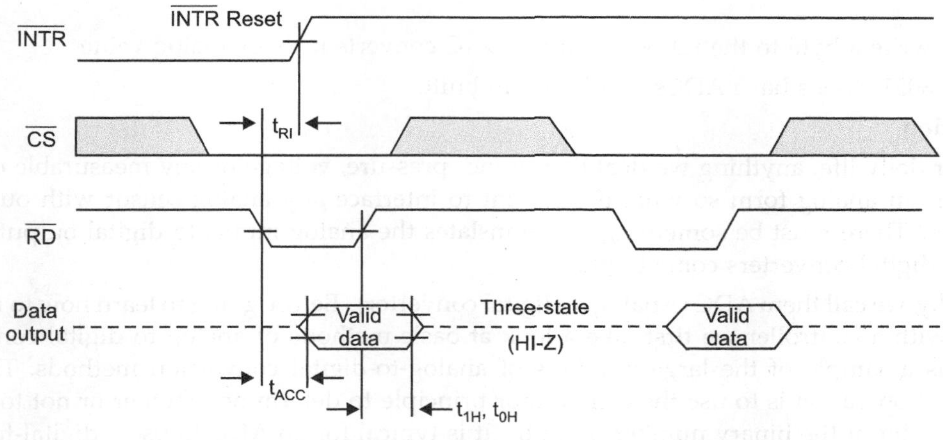

Fig. 9.17 Output Enable and Reset INTR

1. Make chip select (CS) signal low.
2. Make write (WR) signal low.
3. Make chip select (CS) high.
4. Wait for INTR pin to go low (means conversion ends).

Once the conversion in ADC is done, the data is available in the output latch of the ADC. Looking at the Figure 9.17 which shows the timing diagram of how to read the converted value from the output latch of the ADC. Data of the new conversion is only available for reading after ADC0804 made INTR pin low or say when the conversion is over. Below are the stets to read output from the ADC0804.

1. Make chip select (CS) pin low.
2. Make read (RD) signal low.
3. Read the data from port where ADC is connected.
4. Make read (RD) signal high.
5. Make chip select (CS) high.

In the next section of this tutorial we will follow the above mentioned steps to program the ADC.

9.6 DIGITAL TO ANALOG CONVERTER - DAC

Commonly used DAC808 (MC1408)

– R/2R ladder

– Iout = Iref (D7/2 + D6/4 + D5/8 + …… + D0/256)

– Iout converted to voltage by a resistive load or op-amp based isolator (Rf from Vout to V- and V+ to GND)

PinOut

– D0-D7 Connected to the Processor's IO port

– Vref+, Vref-, Vee

Usage:

– Just write a byte to the IO port and the DAC converts it to an analog value.

Some 8051 clones have ADCs and DACs in built.

Introduction

In our daily life, anything we deal like sound, pressure, voltage or any measurable quantity, are usually in analog form so what if we want to interface any analog sensor with our digital controllers? There must be something that translates the analog inputs to digital output, and so analog to digital converters come to play.

Usually, we call them ADC (Analog to digital converter). Before going to learn how to interface an ADC with a controller we first take a look at basic methods of analog to digital conversion. This is a sample of the large numbers of analog-to-digital conversion methods. The basic principle of operation is to use the comparator principle to determine whether or not to turn on a particular bit of the binary number output. It is typical for an ADC to use a digital-to-analog converter (DAC) to determine one of the inputs to the comparator.

Following are the most used conversion methods:

Digital-Ramp ADC

Successive Approximation ADC

Flash ADC

Digital-Ramp ADC

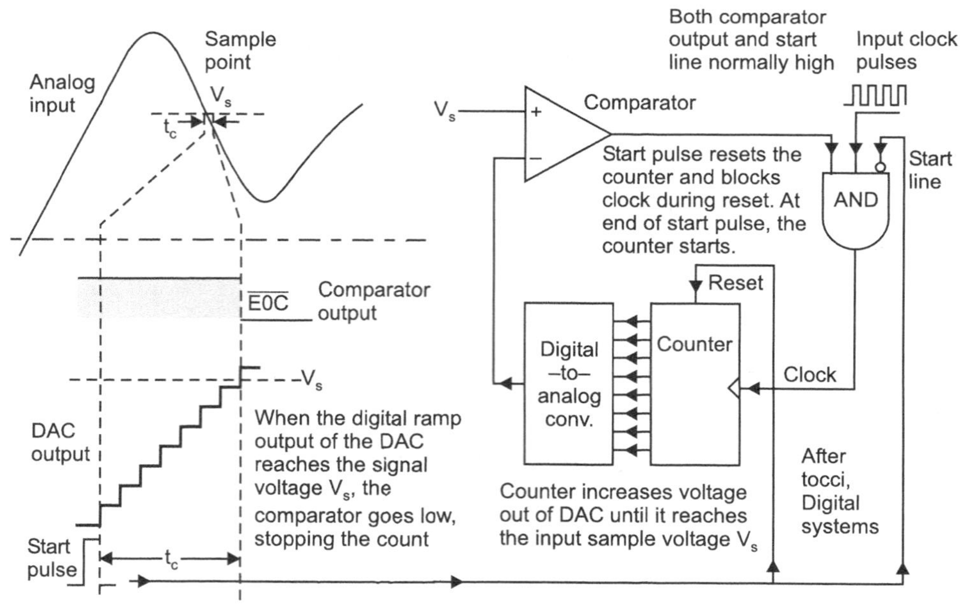

Fig. 9.18 Digital RAM ADC

Conversion from analog to digital form inherently involves comparator action, where the value of the analog voltage at some points in time is compared with some standards. A common way to do that is to apply the analog voltage to one terminal of a comparator and trigger a binary counter which drives a DAC. The output of the DAC is applied to the other terminal of the comparator. Since, the output of the DAC is increasing with the counter, it will trigger the comparator at some points when its voltage exceeds the analog input. The transition of the comparator stops the binary counter, which at that point holds the digital value corresponding to the analog voltage.

9.7 SUMMARY

- The keyboard and display devices are the two main components of microcontroller based system.
- Using them user can give and receive information from the microcontroller based system.
- Instead of BCD to seven-segment decoder (IC 7447) transistors are used to drive the LED segments.
- Due to this we can also display HEX characters on the display.
- However, in this case we have to send the proper 7-segment code of a particular digit that is to be displayed on the port1.
- ADC implies sampling and encoding a continuous time signal whereas DAC is to produce a quantized analogue output corresponding to a particular binary-digital input code.
- The interfacing of ADC 0803/0804/0805 with 8051 using port 1 and port 2.
- Here, port 1 is used to read digital data from ADC and port 2 is used to provide control signals to ADC 0803/0804/0805.
- The conversion time is around 110μs.

9.8 QUESTIONS

1. Traffic light can be implemented with the peripheral

   (*a*) 8255 (*b*) 8253 (*c*) 8259 (*d*) 8357

2. Multiplexing of display is used

   (*a*) to save power (*b*) to increase the speed

   (*c*) to decrease the speed (*d*) none of these

3. In traffic control the number of LED used controlling the vehicle flow is

   (*a*) Two (*b*) Three (*c*) Four (*d*) None

4. The digital to analog converter with

   (*a*) O/P digital data (*b*) I/P analog data

   (*c*) O/P data (*d*) O/P analog data

5. Seven segment display can be interfaced with

   (*a*) 8257 (*b*) 8253 (*c*) 8279 (*d*) 8259

6. The segments of a seven-segment display are lettered to a

   (*a*) Clockwise direction (*b*) Counter clockwise direction

   (*c*) either of (*a*) or (*b*) above

7. Current drawn when the number 8 is on an LED display is

   (*a*) 140 nA (*b*) 140 μA (*c*) 140 mA (*d*) None of these

8. Current supplied to a four digit liquid crystal display that reads the number 8888 is of the order of

   (*a*) 560 nA (*b*) 560 μA (*c*) 560 mA (*d*) 5.6 A

9. How many outputs are there in the output of a 10-bit D/A converter?

   (*a*) 1000 (*b*) 1023 (*c*) 1024 (*d*) 1224

10. What is the normal range of analog input voltage?

    (*a*) 0 to 1V (*b*) 0 to 5V (*c*) 5 to 15V (*d*) 14V to 30V.

11. If V_{IN} is 0.99 V, what is the digital output of the ADC 0801 after INTR goes low?

    (*a*) 0011 0011 (*b*) 0101 1111 (*c*) 0111 1100 (*d*) 1111 1111.

Chapter 10

PROGRAM

10.1 INTRODUCTION

An assembly language program is a set of instruction written in the mnemonics of a given microcontroller. These instructions are the comments to the microcontroller to be executed in the given sequence to accomplish a task to write such programs for the microcontroller. We should be familiar with the programming model and the instruction set of the microcontroller.

10.2 8-BIT ADDITION

Example : 1

```
MOV A, #05            ; Load 05 data into accumulator
MOV R2, #02           ; Load 02 data into R2
ADD A, R2             ; Add accumulator and R2
MOV DPTR, #4000       ; Load the address to DPTR(data pointer)
MOV X @DPTR, A        ; Get the number in the accumulator
                   A = 05H
                  R0 = 02H
                       -----
                       07H
                       ------
```

Example : 2

```
MOV A, #03            ; Load 03 data into accumulator.
ADD A, #07            ; (A) ← (A) + 07H.
MOV DPTR, #4002       ; Load the address to data pointer.
MOV X @DPTR, A        ; Get the number in the accumulator.
                   A = 03H
```

```
Data = 07H
       ------
       0AH
       -------
```

Example : 3

```
MOV DPTR, #4200      ; Load the address to the data pointer.
MOV X A,@DPTR        ; Get the memory location to accumulator.
MOV R0, #08          ; Load 08 data into register R0.
ADD A, R0            ; (A) ← (A) + (R0).
MOV DPTR, #4008      ; Load the address to data pointer.
MOV X @DPTR, A       ; Get the number in the accumulator.
```

10.3 8-BIT SUBTRACTION

```
MOV A, #2A           ; Get the first number in A
MOV R0, #22          ; Get the second number in R0
CLR C                ; Clear carry
SUBB A, R0           ; A ← A – (R0)
SWAP A               ; Exchange digits
          A = 2AH
         R0 = 22H
              ------
          A = 08H
     SWAP A = 80H
```

10.4 16-BIT ADDITION

```
MOV DPTR, #2242H     ; (DPTR) ← 2242H (16-bit number)
MOV A, #2BH          ; (A) ← 2BH (lower bytes of second 16-bit number)
MOV B, #20H          ; (B) ← 20H (higher bytes of second 16-bit number)
ADD A, DPL           ; Add lower bytes
MOV DPL, A           ; Save result of lower bytes addition
MOV A,B              ; Get higher bytes of second number in A
ADDC A, DPH          ; Add higher bytes with any carry from lower bytes addition
MOV DPH, A           ; Save result of higher bytes addition
```

10.5 16 BIT SUBTRACTION

```
MOV A, DATA L1       ; move the lower order data in to a
MOV R, # DATA L2     ; move the higher order 07 in to r1
SUBB A,R1            ; subtract with borrow the content of a, and data rL
MOV DPTR, #4500      ; move the address 4500 in dptr
```

```
MOVX @DPTR, A          ; move the result in a register to external memory whose address
                       is at data pointer.
INC DPTR               ; Inc dptr add 1 to data pointer
MOV A, #M1             ; move high byte data 1 in to a register
SUBB A, # M2           ; subtract with borrow the content of a and high order byte of data
                       2
MOVX @ DPTR, A         ; move the result in a in to memory whose address is at data
                       pointer
SJUMP                  ; Sjump hlt
```

10.6 SUBTRACT TWO 8-BIT NUMBERS AND EXCHANGE DIGITS

```
MOV A, #9F             ; Get the first number in A
MOV R0, #40            ; Get the second number in R0
CLR C                  ; Clear carry
SUBB A, R0             ; A ← A – (R0)
SWAP A                 ; Exchange digits
```

10.7 MULTIPLY TWO 8-BIT NUMBERS

```
MOV A, #07             ; Get the first number in A
MOV B, #02             ; Get the second number in B
MUL AB                 ; A × B, Higher bytes of result in B
                       ; and lower bytes of result in A
MOV DPTR, #4008        ; Load the address to data pointer
MOV X @DPTR, A         ; Get the number in the accumulator
```

Before Execution

A = 07H

B = 02H

After Execution

A = 00H

B = 0EH

10.8 DIVISION TWO 8-BIT NUMBERS

```
MOV A, #0A             ; Get the first number in A
MOV B, #04             ; Get the second number in B
DIV AB                 ; A ÷ B, Remainder in B and Quotient
                       ; in A
MOV DPTR, #4008        ; Load the address to data pointer
MOV X @DPTR, A         ; Get the number in the accumulator
```

Before Execution

A = 0AH

B = 04H

After Execution

A = 02H

B = 02H

10.9 ARITHMETIC AND LOGIC OPERATIONS

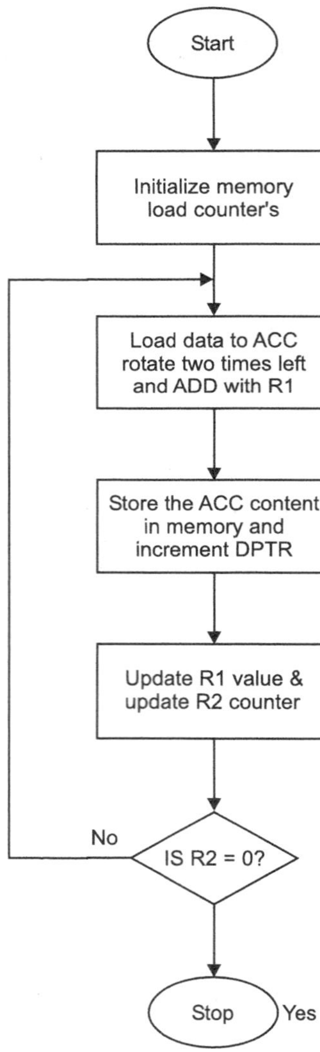


```
START : MOV DPTR, #8050      ; DPTR = 8050
MOV R1, #01H                 ; 01 → R1
MOV R2, #0AH                 ; 0A → R2
LOOP : MOV A,R1              ; R1 → A
```

```
RL A                    ; ROTATE LEFT
RL A                    ; ROTATE  LEFT
ADD A, R1               ; A = A + R1
MOVX @ DPTR, A          ; A → 8050H
INC DPTR                ; DPTR = DPTR + 1
INC R1                  ; R1 = R1 + 1
DJNZ R2, 8007           ; R2 = R2 – 1, (Z=0) JUMP
```

Output :

8050 – 05H
8051 – 0AH
8052 – 0FH
8053 – 14H
8054 – 19H
8055 – 1EH
8056 – 23H
8057 – 28H
8058 – 2DH
8059 – 32H

10.10 UP/DOWN COUNTER AND OBJECT COUNTER

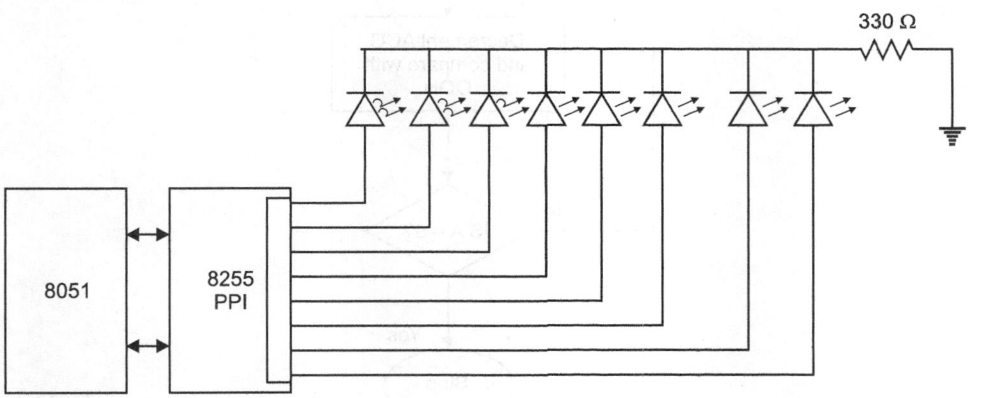

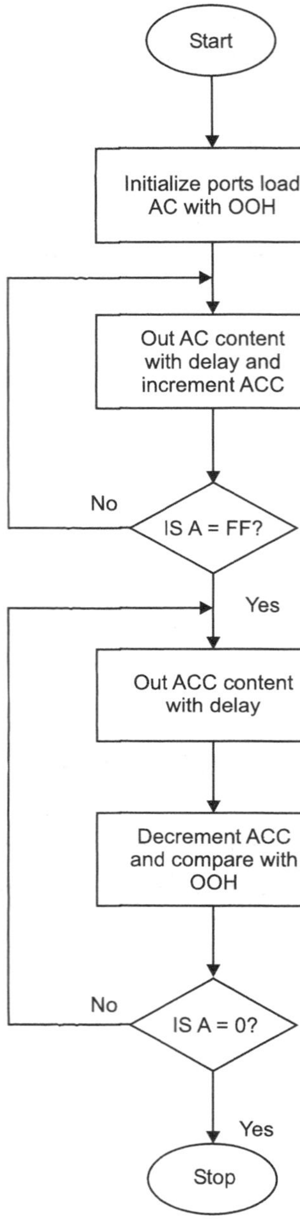


```
START :   MOV DPTR, #2023       ; 2023 = DPTR
          MOV A, #80            ; A = 80
          MOVX @ DPTR, A        ; A → C.R
          MOV DPTR, #2020       ; DPTR = 2020
          MOV A, #00            ; A= 00
LOOP :    MOVX @ DPTR, A        ; A → P.A
          ACALL DELAY           ; CALL DELAY
```

```
            INC A                   ; A = A + 1
            CJNE A, #OF, 800B       ; A (CMP) OF & JUMP ≠
            MOVX @ DPTR, A          ; A → P.A
            ACALL DELAY             ; CALL DELAY
            DEC A                   ; A = A – 1
            CJNE A, #00, 8012       ; A (CMP) 00 & JMP ≠
            MOVX @ DPTR, A          ; A → P.A
DELAY       MOV R3, #04             ; 04 → R3
LOOP 2 :    MOV R1, #FF             ; FF → R1
LOOP 1 :    MOV R2, #FF             ; FF → R2
LOOP :      DJNZ R2, LOOP           ; R2 = R2 – 1 & JMP IF Z = 0
            DINZ R1, LOOP1          ; R1 = R1 – 1 & JMP IF Z = 0
            DINZ R3, LOOP2          ; R3 = R3 – 1 & JMP IF Z = 0
            RET                     ; RETURN
```

OBJECT COUNTER

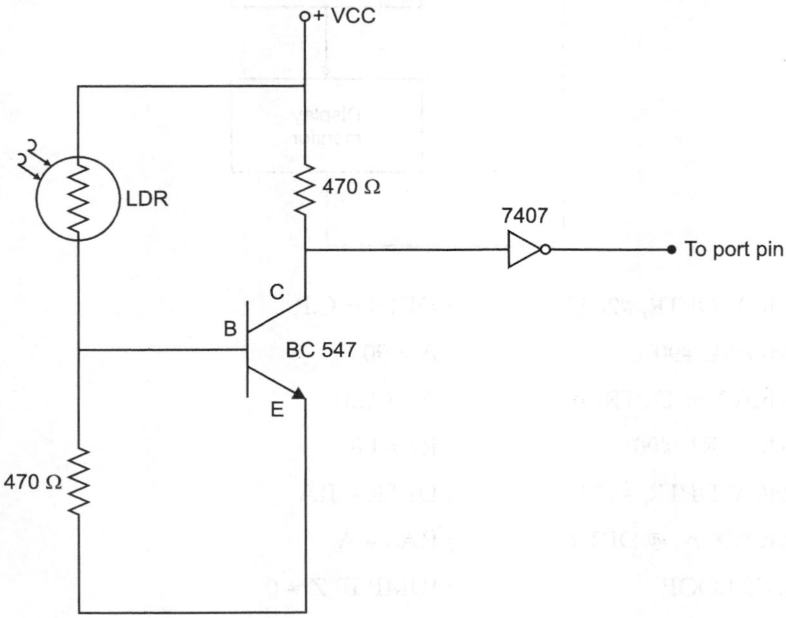

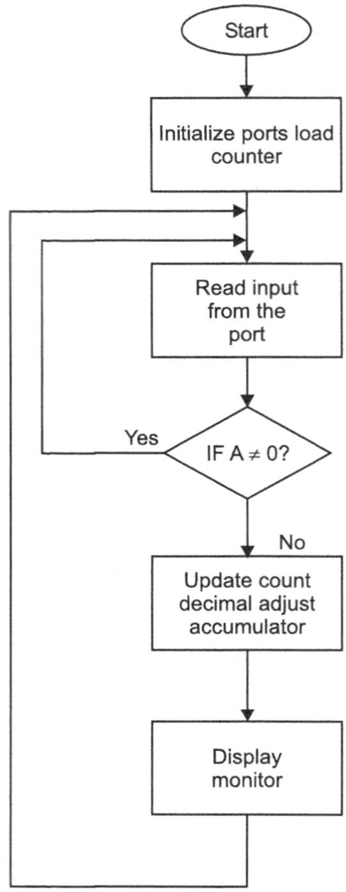


```
START :  MOV DPTR, #2023      ; DPTR = C.R
         MOVA, #90            ; A = 90
         MOVX @ DPTR, A       ; A → C.R
         MOV R1, #00          ; R1 = 00
LOOP :   MOV DPTR, #2020      ; DPTR = P.A
         MOVX A, @ DPTR       ; P.A → A
         JNZ LOOP             ; JUMP IF Z ≠ 0
         MOV A, R1            ; R1 → A
         ADD A, #01           ; A = A + 01
         DA A                 ; Decimal Adjust Accumulator
         MOV R1, A            ; A → R1
         MOV R6, A            ; A → R6
         LCALL 677D           ; CALL DISPLAY ROUTINE
```

```
         LCALL                  ; CALL DELAY
         LJMP LOOP              ; JUMP TO LOOP
DELAY    MOV R3, #04            ; 04 → R3
LOOP 2 : MOV R1, #FF            ; FF → R1
LOOP 1 : MOV R2, #FF            ; FF → R2
LOOP :   DJNZ R2, LOOP          ; R2 = R2 – 1 & JMP IF Z = 0
         DINZ R1, LOOP1         ; R1 = R1 – 1 & JMP IF Z = 0
         DINZ R3, LOOP2         ; R3 = R3 – 1 & JMP IF Z = 0
         RET                    ; RETURN
```

10.11 ANALOG TO DIGITAL CONVERTER

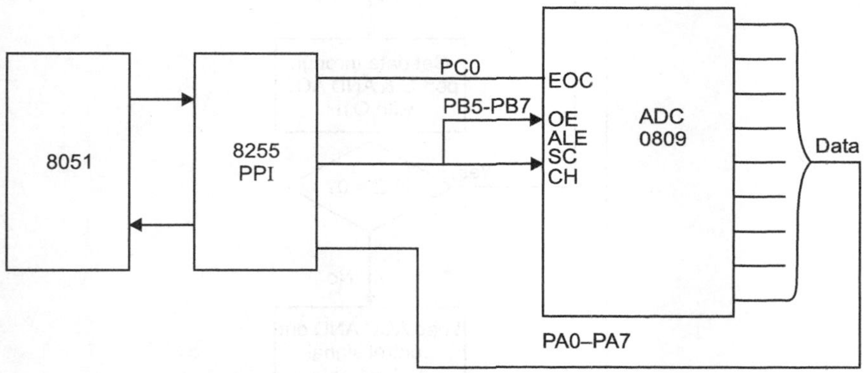

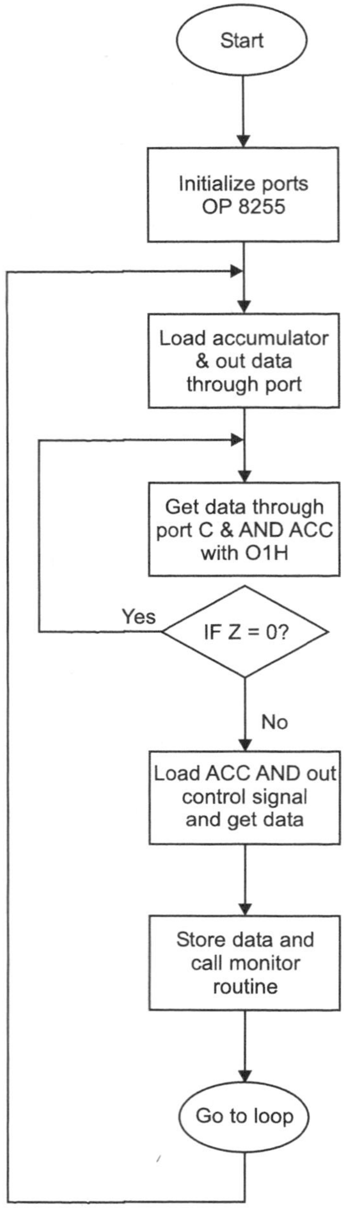


```
START :  MOV DPTR, #2023        ; 2023 → DPTR
         MOVA A, #99            ; A = #99
         MOVX @ DPTR, A         ; A → C.R
         LOOP MOV A, #80        ; A = #80
         MOV DPTR, #2021        ; DPTR = P.B
         MOVX @ DPTR, A         ; A = P.B
         MOV A, #E0             ; A = #E0
         MOVX @ DPTR,  A        ; A = P.B
```

```
LOOP 1 : MOV DPTR, #2022        ; DPTR = P.C
         MOVX A, @ DPTR         ; P.C → A
         ANL A, #01             ; A^ #01 = A
         JNZ 900F               ; JUMP IF Z = 0
         MOV A, #60             ; A = #60
         MOV DPTR, # 2021       ; DPTR = P.B
         MOVX @ DPTR, A         ; A → P.B
         MOV DPTR, #2020        ; DPTR = P.A
         MOVX A, @DPTR          ; P.A → A
         MOV R6, A              ; A → R6
         LCALL 677D             ; CALL Monitor Routine
         LIMP LOOP              ; Imp to Loop
```

ANALOG INPUT	DIGITAL OUTPUT
0.5	1D
1.0	37
2.0	69
2.5	81
3.0	90
4.0	D0
5.0	FF

10.12 DATA TRANSFER WITH PARALLEL PORTS

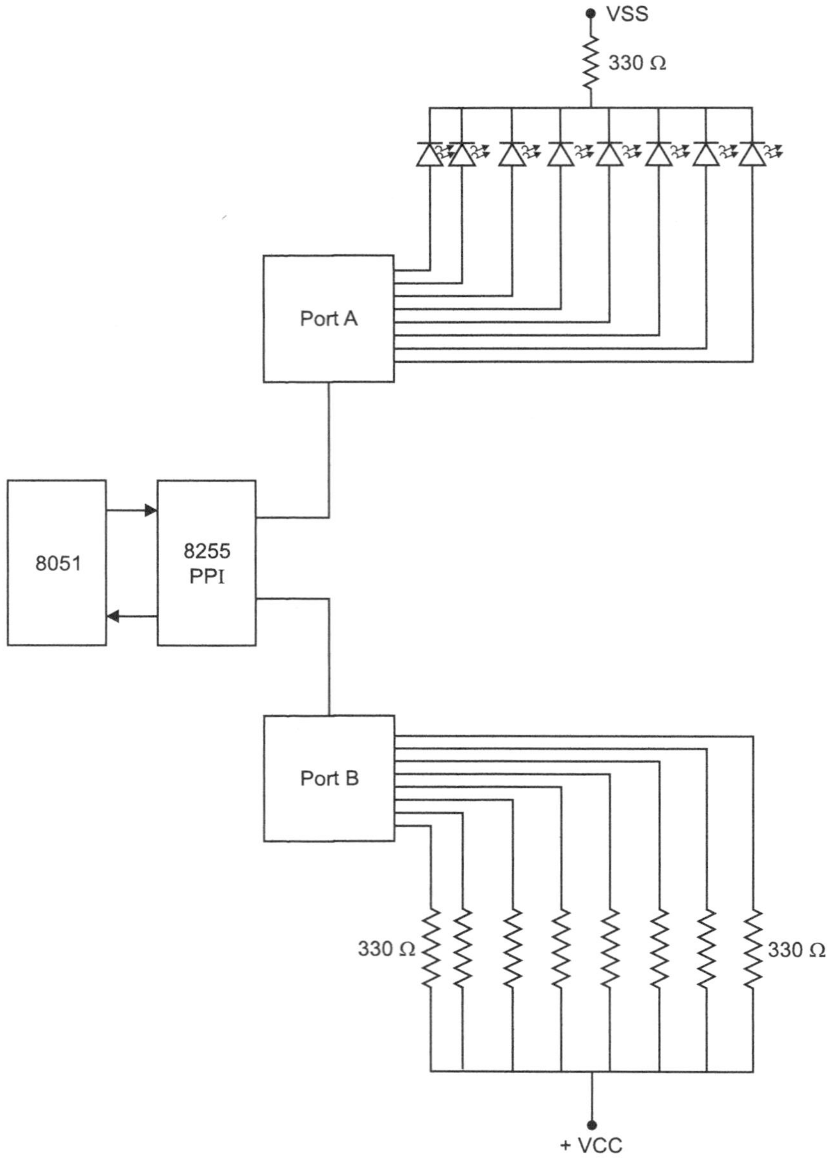

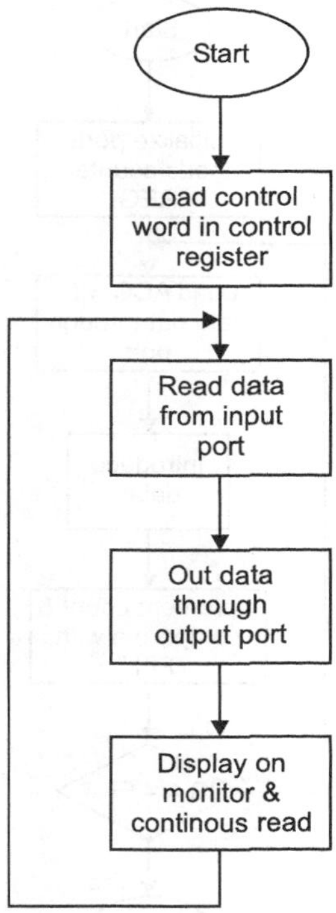


```
START :  MOV DPTR, #2023      ; DPTR = 2023
         MOV A, #82           ; A = 82
         MOVX @ DPTR, A       ; A → DPTR
         MOV DPTR, #2021      ; DPTR = P.B
         MOVX A, @ DPTR       ; P.A → A
         MOV R1, A            ; A → R1
         MOV DPTR, #2020      ; DPTR → P.A
         MOVX @ DPTR, A       ; A → P.A
         MOV A, R1            ; R1 → A
         MOV R6, A            ; A → R6
         LCALL 677D           ; CALL DISPLAY
         LJMP B006            ; JMP TO LOOP
```

10.13 DIGITAL TO ANALOG CONVERTER

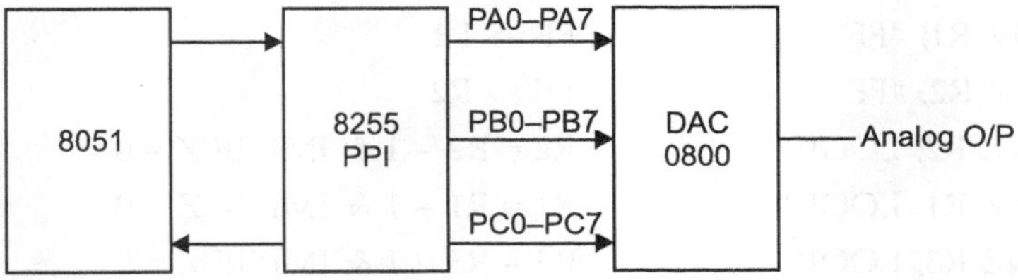

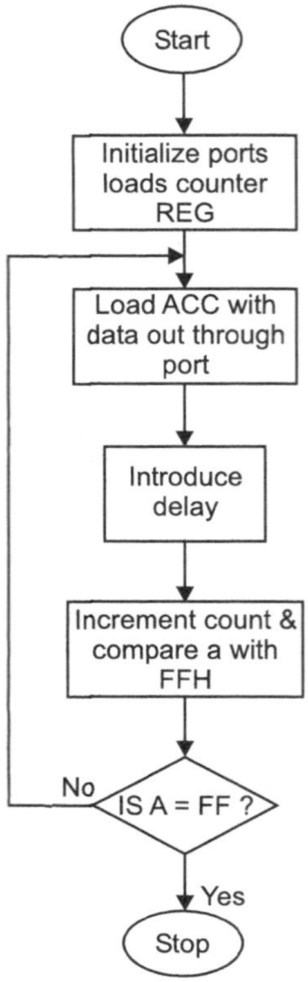


```
START :   MOV DPTR, #2023       ; DPTR = C.R
          MOV A, #80            ; AC = 80
          MOVX @ DPTR, A        ; A → C.R
          MOV R1, #00           ; R1 = 00
LOOP      MOV DPTR, #2020       ; DPTR = P.A
          MOV A, R1             ; R1 → AC
          MOVX @ DPTR, A        ; AC → P.A
          MOV R6, A             ; A → R6
          LCALL 677D            ; CALL DISPLAY ROUTINE
          LCALL DELAY           ; CALL DELAY
          INC R1                ; R1 = R1 + 1
          CJNE R1, #FF, C008    ; COMPARE R1 & #FF, JMP ≠
DELAY     MOV R3, #04           ; 04 → R3
LOOP 2 :  MOV R1, #FF           ; FF → R1
LOOP 1 :  MOV R2, #FF           ; FF → R2
LOOP :    DJNZ R2, LOOP         ; R2 = R2 – 1 & JMP IF Z = 0
          DINZ R1, LOOP1        ; R1 = R1 – 1 & JMP IF Z = 0
          DINZ R3, LOOP2        ; R3 = R3 – 1 & JMP IF Z = 0
          RET                   ; RETURN
```

DIGITAL INPUT	ANALOG OUTPUT
3A	1.13
7B	2.38
AD	3.34
B9	3.57
D6	4.13
FF	4.92

10.14 STEPPER MOTOR INTERFACE

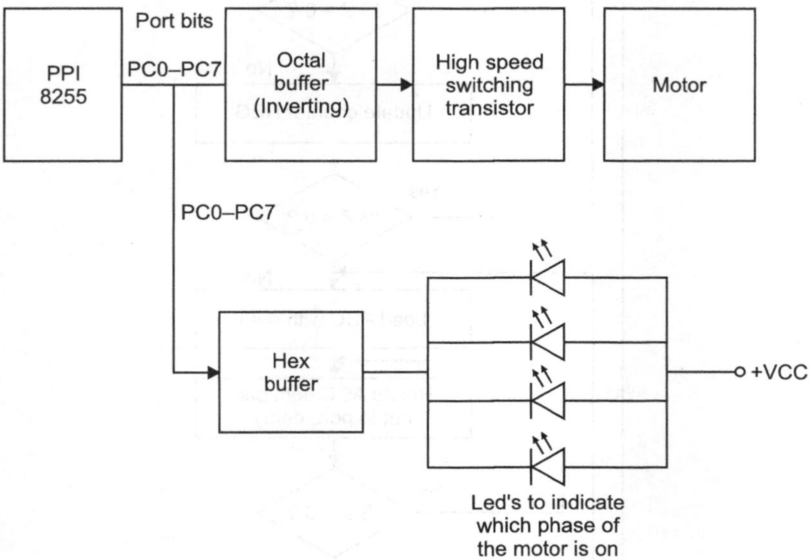

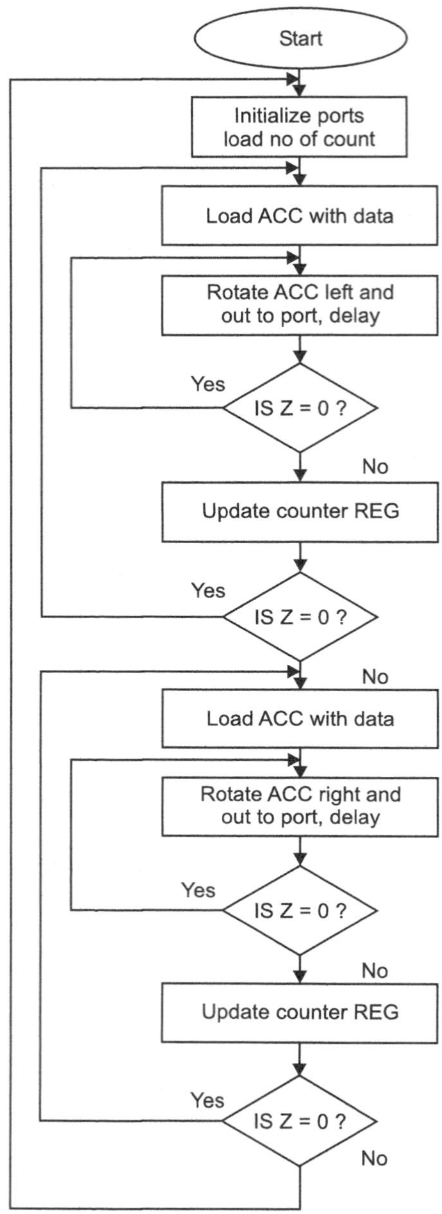


```
START :   MOV DPTR, #2023      ; DPTR = C.R
          MOV A, #80           ; A = 80
          MOVX @ DPTR, A       ; A → C.R
          MOV R1, #19          ; R1 = 19H
LOOP :    MOV A, #77           ; A = 77H
LOOP 1 :  RL A                 ; ROTATE ACCUMULATOR LEFT
          MOV DPTR, #2022      ; DPTR = P.C
```

```
            MOVX @ DPTR, A          ; A → P.C
            LCALL DELAY             ; CALL DELAY
            CJNE A, #77, LOOP1      ; A CMP #77, IF ≠ JMP
            DJNZ R1, LOOP           ; R1 = R1 – 1 & ≠ 0 JMP
            MOV R1, #19             ; R1 = 19H
LOOP 3 :    MOV A, #EE              ; A = EEH
LOOP 4 :    RR A                    ; ROTATE AC RIGHT
            MOV DPTR, #2022         ; DPTR = P.C
            MOVX @ DPTR, A          ; A → P.C
            LCALL DELAY             ; CALL DELAY
            CINE A, #EE, LOOP2      ; A CMP #EE, IF ≠ JMP
            DINZ R1, LOOP3          ; R1 = R1 – 1 & ≠ 0 JMP
DELAY       MOV R3, #04             ; 04 → R3
LOOP 2 :    MOV R1, #FF             ; FF → R1
LOOP 1 :    MOV R2, #FF             ; FF → R2
LOOP :      DJNZ R2, LOOP           ; R2 = R2 – 1 & JMP IF Z = 0
            DINZ R1, LOOP1          ; R1 = R1 – 1 & JMP IF Z = 0
            DINZ R3, LOOP2          ; R3 = R3 – 1 & JMP IF Z = 0
            RET                     ; RETURN
```

10.15 MATRIX KEYPAD AND SSD INTERFACE

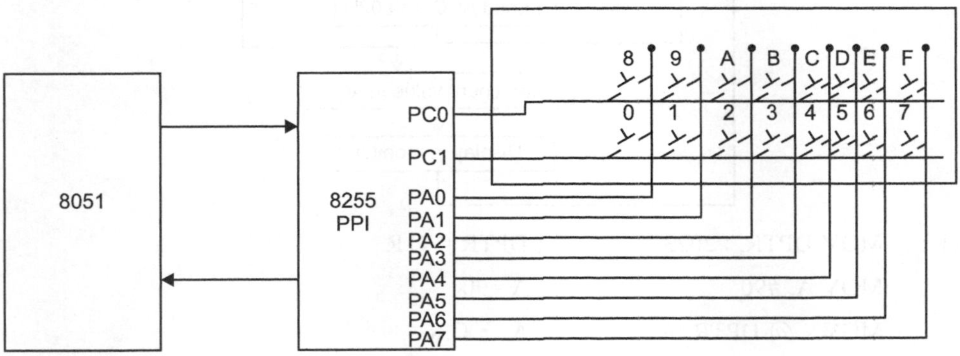

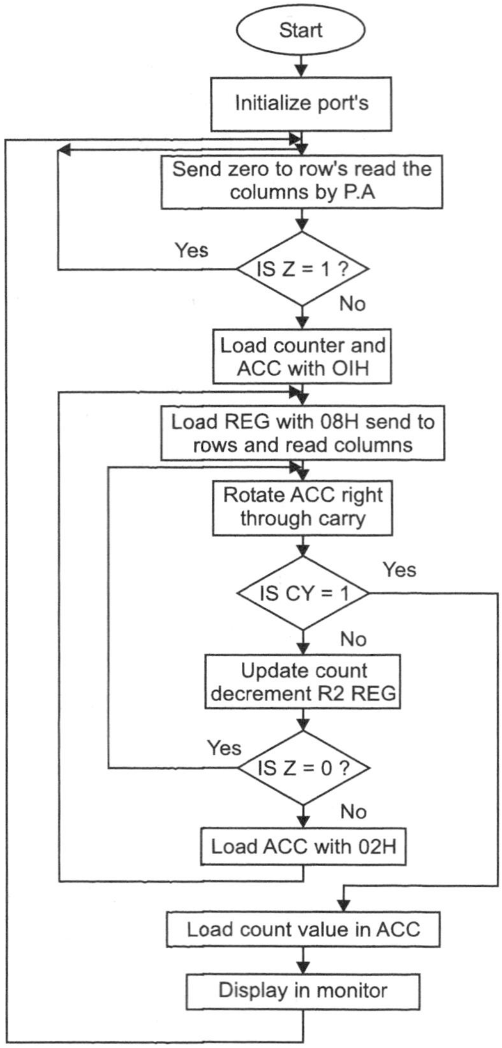


```
START :   MOV DPTR, #2023        ; DPTR = C.R
          MOV A, #90             ; A =90
          MOVX @ DPTR, A         ; A → C.R
LOOP 1 :  MOV DPTR, #2022        ; DPTR = P.C
          MOV A, #D3             ; A = D3
          MOVX @ DPTR, A         ; A → P.C
          MOV DPTR, #2020        ; DPTR = P.A
          MOVX A, @ DPTR         ; P.A → A
          JZ LOOP 1              ; JUMP ON ZERO
          MOV R1, #00            ; R1 = 00
```

```
            MOV A, #01            ; A = 01
LOOP 2 :    MOV R2, #08           ; R2 = 08
            MOV DPTR, #2022       ; DPTR = P.C
            MOVX @ DPTR,A         ; A→ P.C
            MOV DPTR, #2020       ; DPTR = P.A
            MOVX A, @ DPTR        ; P.A → A
LOOP 3 :    RRC A                 ; ROTATE AC RIGHT WITH C
            JC LOOP 4             ; JUMP IF C = 1
            INC R1                ; R1= R1 + 1
            DJNZ R2, LOOP3        ; R2 = R2 – 1, ≠ 0 JMP
            MOV A, # 02           ; A = # 02 H
            LJMP LOOP2             JUMP TO LOOP
LOOP 4 :    MOV A, R1             ; R1 → A
            MOV R6,A              ; A → R6
            LCALL 677D            ; CALL MONITOR ROUTINE
            LJMP LOOP1            ; JUMP TO LOOP1
```

10.16 DIGITAL CLOCK

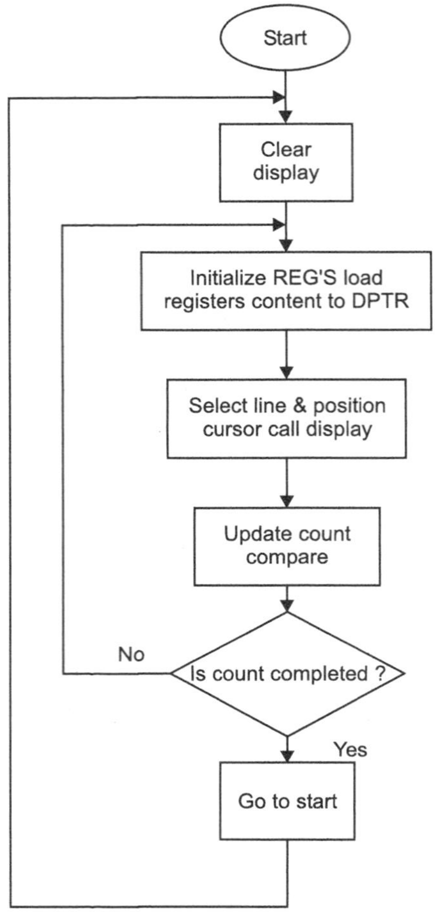


```
START :   LCALL 68EA            ; CLEAR DISPLAY
          MOV R1, #30           ; R1 = #30
LOOP 6 :  MOV A, R1             ; R1 → A
          MOV DPTR, #8164       ; DPTR = 8164
          MOVX @ DPTR, A        ; A → 8164
          MOV R2, #30           ; R2 = #30
LOOP 5 :  MOV A, R2             ; R2 → A
          MOV DPTR, #8165       ; DPTR = 8165
          MOVX @ DPTR, A        ; A → 8165
          MOV R3, #30           ; R3 = #30
LOOP 4 :  MOV A, R3             ; R3 → A
          MOV DPTR, #8167       ; DPTR = 8167
          MOV X @ DPTR, A       ; A = 8167
          MOV R4, #30           ;  R4 = #30
```

```
LOOP 3 :   MOV A, R4                 ; R4 → A
           MOV DPTR, #8168           ; DPTR = 8168
           MOV X @ DPTR, A           ; A = 8168
           MOV R5, #30               ;  R5 = #30
LOOP 2 :   MOV A, R5                 ; R5 → A
           MOV DPTR, #816A           ; DPTR = 816A
           MOV X @ DPTR, A           ; A = 816A
           MOV R6, #30               ;  R6 = #30
LOOP 1 :   MOV A,R6                  ; R6 → A
           MOV DPTR, #816B           ; DPTR = 816B
           MOV X @ DPTR, A           ; A = 816B
           MOV 0F0, #30              ;  0F0 = #30
           LCALL 6946                ; POSITION CURSOR
           MOV DPTR, #8150           ; DPTR = 8150
           LCALL 6919                ; CALL DISPLAY
           LCALL 9400                ; CALL DELAY
           INC R6                    ; R6 = R6 + 1
           CJNE R6, #3A LOOP1        ; R6 CMP #3A & JMP IF ≠
           INC R5                    ; R5 = R5 + 1
           CJNE R5, #36 LOOP2        ; R5 CMP #36 & JMP IF ≠
           INC R4                    ; R4 = R4 + 1
           CJNE R4, #3A LOOP3        ; R4 CMP #3A & JMP IF ≠
           INC R3                    ; R3 = R3 + 1
           CJNE R3, #36 LOOP4        ; R3 CMP #36 & JMP IF ≠
           MOV DPTR, # 8164          ; DPTR = 8164
           MOVX A, @ DPTR            ; 8164 → A
           MOV R1, A                 ; A → R1
           CJNE R1, #32 LOOP7        ; R1 CMP #32 & JMP IF ≠
           INC R2                    ; R2 = R2 + 1
           CJNE R2, #34 LOOP5        ; R2 CMP #34 & JMP IF ≠
           LJMP START                ; JUMP TO START
LOOP 7 :   INC R2                    ; R2 = R2 + 1
           CINE R2, #3A, LOOP5       ; R2 CMP #3A & JMP IF
           CJNE R4, #3A LOOP3        ; R4 CMP #3A & JMP IF ≠
           INC R3                    ; R3 = R3 + 1
           CJNE R3, #36 LOOP4        ; R3 CMP #36 & JMP IF ≠
           INC R1                    ; R1 = R1 + 1
           LJMP LOOP6                ; JUMP TO LOOP 6
```

```
DELAY :    MOV R0, #05          ; R0 = 05
LOOP 2 :   MOV R7, #FF          ; R7 = FF
LOOP1 ;    MOV DPTR, #9600      ; DPTR = 9600
           MOVX A, @ DPTR       ; 9600 → A
           DEC A                ; A = A – 1
           INZ LOOP             ; JUMP ON ≠ 0
           DJNZ R7, LOOP1       ; R7 = R7 – 1, ≠ 0 JMP
           DJNZ R0, LOOP2       ; R0 = R0 – 1, ≠ 0 JMP
           RET                  ; RETURN
```

APPENDIX A

SERIAL COMMUNICATION

RS-232 WAVEFORM

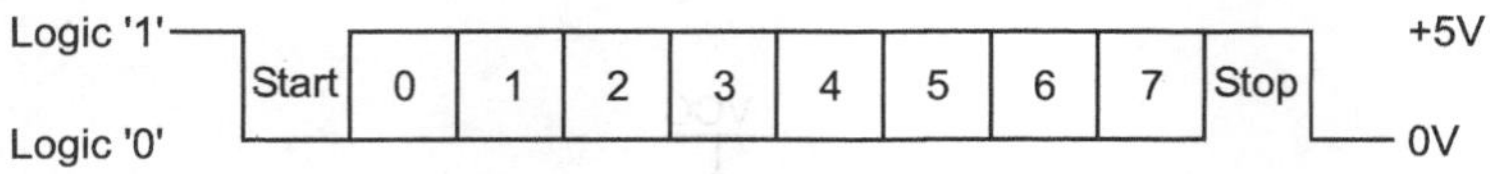

TTL/CMOS SERIAL LOGIC WAVEFORM

The diagram above shows the expected waveform from the UART when using the common 8N1 format. 8N1 signifies 8 Data bits, No Parity and 1 Stop Bit. The RS-232 line, when idle is in the Mark State (Logic 1). A transmission starts with a start bit which is (Logic 0). Then each bit is sent down the line, one at a time.

The LSB (Least Significant Bit) is sent first. A Stop Bit (Logic 1) is then appended to the signal to make up the transmission. The data sent using this method, is said to be *framed*. That is the data is *framed* between a Start and Stop Bit.

RS-232 Voltage levels

1. +3 to +25 volts to signify a "Space" (Logic 0).
2. –3 to -25 volts for a "Mark" (logic 1).
3. Any voltage in between these regions (i.e. between +3 and –3 Volts) is undefined.

The data byte is always transmitted *least-significant-bit first*.

The bits are transmitted at specific time intervals determined by the **baud rate** of the serial signal.

This is the signal present on the RS-232 Port of your computer, shown below.

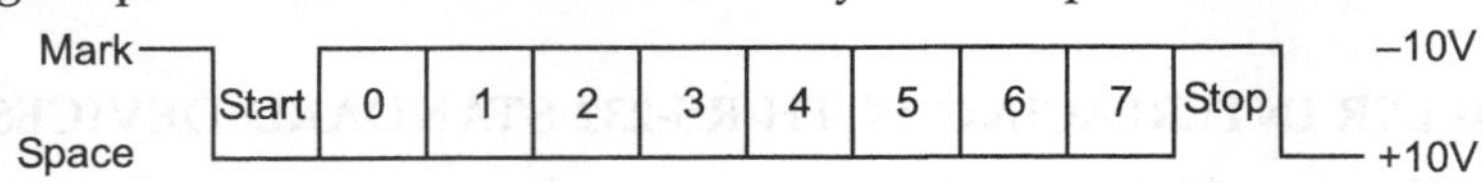

RS-232 Logic Waveform

RS-232 LEVEL CONVERTER

Standard serial interfacing of microcontroller (TTL) with PC or any RS232C Standard device, requires TTL to RS232 Level converter . A MAX232 is used for this purpose. It provides 2-channel RS232C port and requires external 10uF capacitors.

The driver requires a single supply of +5V.

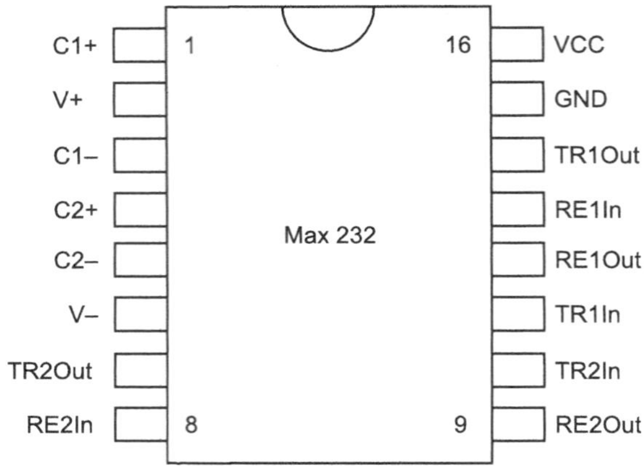

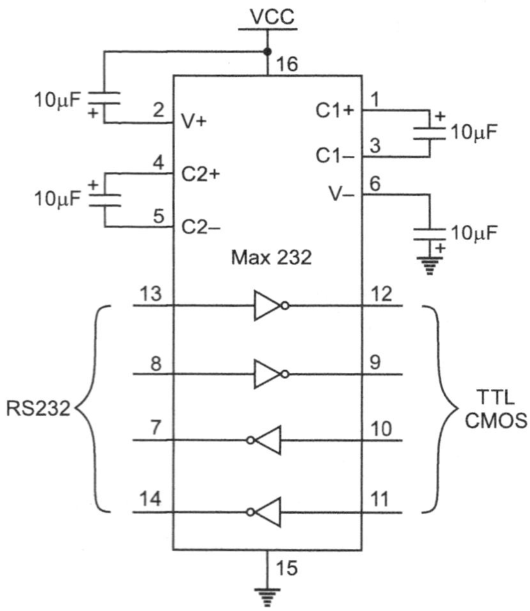

MAX-232 includes a Charge Pump, which generates +10V and –10V from a single 5V supply.

MICROCONTROLLER INTERFACING WITH RS-232 STANDARD DEVICES

- MAX232 (+5V –> + –12V converter)
- Serial port male 9 pin connector (SER)

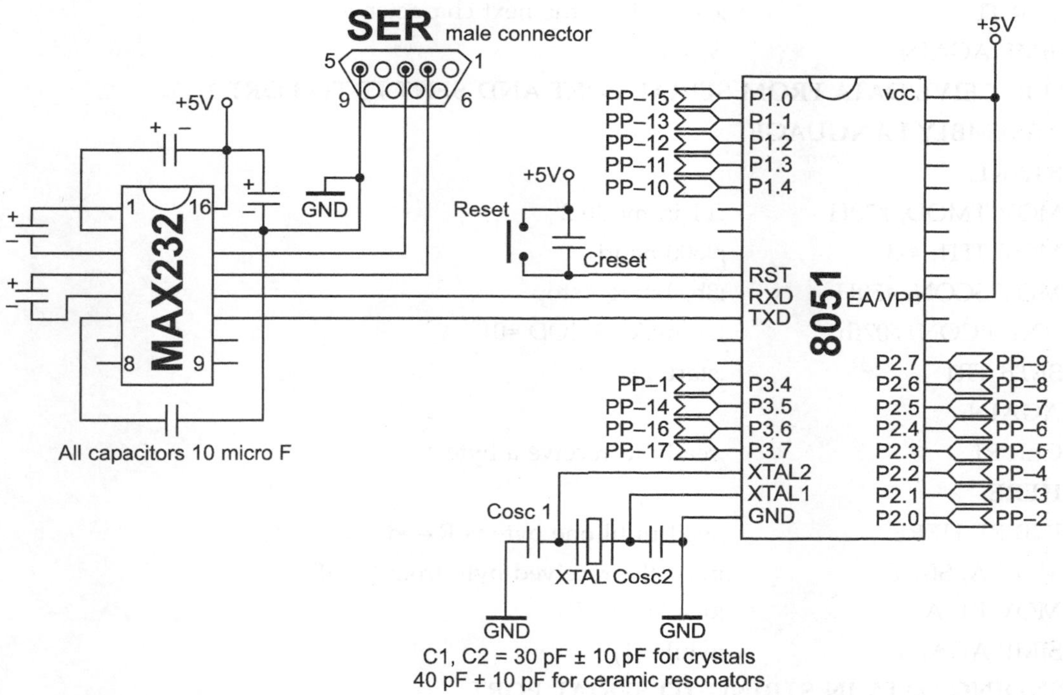

SETTING SERIAL PORT

SCON

8 bit UART, RN enabled, TI & RI operated by program. - 50hex

Timer 1 Count

TH1 = 256 – ((Crystal / 384) / Baud) -PCON.7 is clear.

TH1 = 256 – ((Crystal / 192) / Baud)-PCON.7 is set.

so with PCON.7 is clear we get timer value = FDhex

CODE EXAMPLE

1. TRANSMITTING 'A' CONTINUOUSLY ON SERIAL PORT

ASSEMBLY LANGUAGE

```
START
MOV TMOD, #20H          ;T1 is mode2
MOV TH1, #0fd           ;9600 baud
MOV SCON, #50H          ;8b, 1stop, 1start, REN enabled
ANL PCON, #07fh         ;To make SMOD = 0
SETB TR1                ; start T1
AGAIN
MOV SBUF, #'A'          ; letter A is transmitted
HERE
JNB TI, HERE            ;poll TI until all the bits are transmitted
```

```
CLR TI                  ;clear TI for the next character
SJMP AGAIN              ;while(1)
```

2. TO RECEIVE DATA FROM SERIAL PORT AND SENT IT TO PORT 1

ASSEMBLY LANGUAGE

```
START:
MOV TMOD, #20H          ;T1 in mode 2
MOV TH1, #-3            ;9600 baud
MOV SCON, #50H          ;8b, 1start, 1stop
ANL PCON, #07fh         ;To make SMOD =0
SETB TR1                ;start T1
AGAIN:
CLR RI                  ;ready to receive a byte
HERE:
JNB RI, HERE            ;wait until one byte is Rx-ed
MOV A, SBUF             ;read the received byte from SBUF
MOV P1, A               ;display on P1
SJMP AGAIN              ;while (1)
```

3. SENDING DATA IN STRING TO SERIAL PORT

In Assembly Lan. prog. : Data is stored in string at pointer **DATA. 0** is appended at end of string. In transmit subroutine data in string is transmitted till 0 is detected.

ASSEMBLY LANGUAGE (A51)

```
.org 0000h
LJMP START
DATA: .db "HI, I AM MAHESH", 0dh, 0ah, 0; 0 at end to detect end of string (0d carrage return, 0a -line feed)
TRANSMIT:
CLR A                   ; clear A to get data
MOVC A,@A+DPTR          ; get data from string at data pointer
JZ EXITSTR              ; if data zero, eos
LCALL OUTCHAR           ; else send character
inc DPTR                ; increment data pointer
SJMP TRANSMIT           ; continue, zero condition will terminate
EXITSTR:
ret
OUTCHAR:
MOV SBUF, A             ; place A into Serial Port 1 Buffer
WAITCHAR:
JNB TI,WAITCHAR         ; wait buffer empty flag is set
```

```
CLR TI                  ; clear buffer empty flag
ret
START:
INITIALISATION
MOV TMOD, #20H          ;T1 in mode 2
MOV TH1, #-3            ;9600 baud
MOV SCON, #50H          ;8b, 1start, 1stop
ANL PCON, #07fh         ;To make SMOD =0
SETB TR1                ;start T1
TO SEND DATA
MOV DPTR, #DATA
LCALL TRANSMIT
SJMP START
```

EXAMPLE - MOBILE PHONE AND GPS RECEIVER

You can use same circuit for communicating with Mobile phones/GSM Module or GPS. Communicating with both of these require a Multiplexer, which can be implemented using NAND gates.

GPS SERIAL OUTPUT

Most GPS are capable of sending information through a simple serial link. Only the TXD and GROUND pins need to be connected. The GPS must be set at 9600 bps (or 4800), 8 bits, No Parity, and 1 stop bit.

AND gate as 2:1 Mux. Which connects Rx of GSM modem or GPS receiver according to select bit logic level (pin P1.0 of uC).

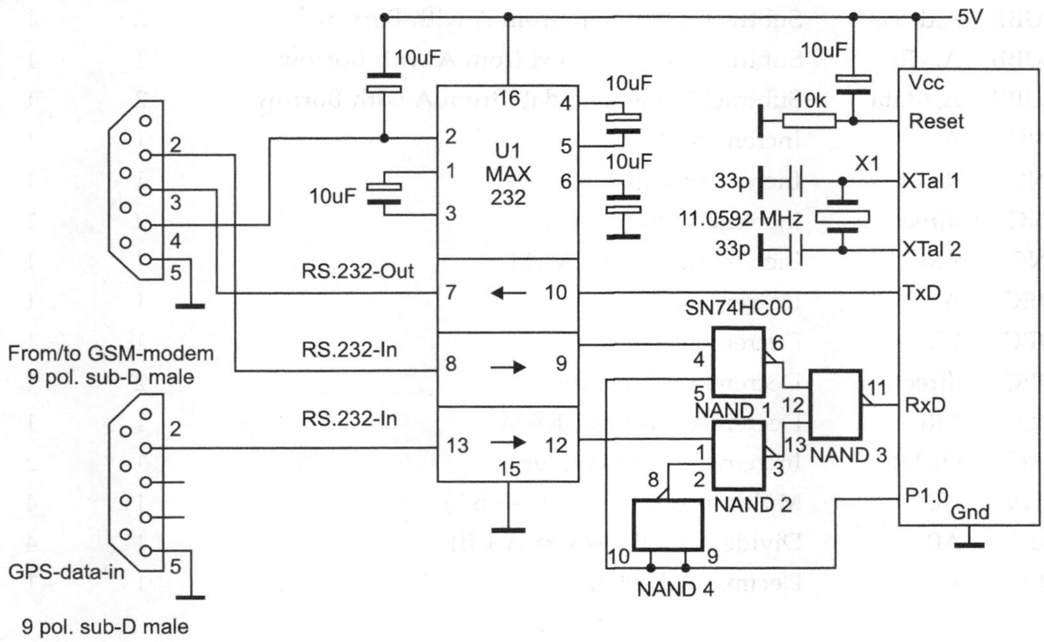

APPENDIX B

INSTRUCTION SET SUMMARY

ARITHMETIC OPERATIONS

Mnemonic		Description	Bytes	Cycles
ADD	A,Rn	Add register to A	1	1
ADD	A,direct	Add direct byte to A	2	1
ADD	A,@Ri	Add indirect RAM to A	1	1
ADD	A,#data	Add immediate data to A	2	1
ADDC	A,Rn	Add register to A with Carry	1	1
ADDC	A,direct	Add direct byte to A with Carry	2	1
ADDC	A,@Ri	Add indirect RAM to A with Carry	1	1
ADDC	A,#data	Add immediate data to A with Carry	2	1
SUBB	A,Rn	Subtract register from A with Borrow	1	1
SUBB	A,direct	Subtract direct byte from A with Borrow	2	1
SUBB	A,@Ri	Subtract indirect RAM from A with Borrow	1	1
SUBB	A,#data	Subtract immediate data from A with Borrow	2	1
INC	A	Increment A	1	1
INC	Rn	Increment register	1	1
INC	direct	Increment direct byte	2	1
INC	@Ri	Increment indirect RAM	1	1
DEC	A	Decrement A	1	1
DEC	Rn	Decrement register	1	1
DEC	direct	Decrement direct byte	2	1
DEC	@Ri	Decrement indirect RAM	1	1
INC	DPTR	Increment Data Pointer	1	2
MUL	AB	Multiply A & B (A × B => BA)	1	4
DIV	AB	Divide A by B (A/B => A + B)	1	4
DA	A	Decimal Adjust A	1	1

LOGICAL OPERATIONS

Mnemonic		Description	Bytes	Cycles
ANL	A,Rn	AND register to A	1	1
ANL	A,direct	AND direct byte to A	2	1
ANL	A,@Ri	AND indirect RAM to A	1	1
ANL	A,#data	AND immediate data to A	2	1
ANL	direct,A	AND A to direct byte	2	1
ANL	direct,#data	AND immediate data to direct byte	3	2
ORL	A,Rn	OR register to A	1	1
ORL	A,direct	OR direct byte to A	2	1
ORL	A,@Ri	OR indirect RAM to A	1	1
ORL	A,#data	OR immediate data to A	2	1
ORL	direct,A	OR A to direct byte	2	1
ORL	direct,#data	OR immediate data to direct byte	3	2
XRL	A,Rn	Exclusive-OR register to A	1	1
XRL	A,direct	Exclusive-OR direct byte to A	2	1
XRL	A,@Ri	Exclusive-OR indirect RAM to A	1	1
XRL	A,#data	Exclusive-OR immediate data to A	2	1
XRL	direct,A	Exclusive-OR A to direct byte	2	1
XRL	direct,#data	Exclusive-OR immediate data to direct byte	3	2
CLR	A	Clear A	1	1
CPL	A	Complement A	1	1
RL	A	Rotate A Left	1	1
RLC	A	Rotate A Left through Carry	1	1
RR	A	Rotate A Right	1	1
RRC	A	Rotate A Right through Carry	1	1
SWAP	A	Swap nibbles within A	1	1

DATA TRANSFER

Mnemonic		Description	Bytes	Cycles
MOV	A,Rn	Move register to A	1	1
MOV	A,direct	Move direct byte to A	2	1
MOV	A,@Ri	Move indirect RAM to A	1	1
MOV	A,#data	Move immediate data to A	2	1
MOV	Rn,A	Move A to register	1	1
MOV	Rn,direct	Move direct byte to register	2	2
MOV	Rn,#data	Move immediate data to register	2	1
MOV	direct,A	Move A to direct byte	2	1

Mnemonic		Description	Bytes	Cycles
MOV	direct,Rn	Move register to direct byte	2	2
MOV	direct,direct	Move direct byte to direct byte	3	2
MOV	direct,@Ri	Move indirect RAM to direct byte	2	2
MOV	direct,#data	Move immediate data to direct byte	3	2
MOV	@Ri,A	Move A to indirect RAM	1	1
MOV	@Ri,direct	Move direct byte to indirect RAM	2	2
MOV	@Ri,#data	Move immediate data to indirect RAM	2	1
MOV	DPTR,#data16	Load Data Pointer with 16-bit constant	2	1
MOVC	A,@A+DPTR	Move Code byte relative to DPTR to A	1	2
MOVC	A,@A+PC	Move Code byte relative to PC to A	1	2
MOVX	A,@Ri	Move External RAM (8-bit addr) to A	1	2
MOVX	A,@DPTR	Move External RAM (16-bit addr) to A	1	2
MOVX	@Ri,A	Move A to External RAM (8-bit addr)	1	2
MOVX	@DPTR,A	Move A to External RAM (16-bit addr)	1	2
PUSH	direct	Push direct byte onto stack	2	2
POP	direct	Pop direct byte from stack	2	2
XCH	A,Rn	Exchange register with A	1	1
XCH	A,direct	Exchange direct byte with A	2	1
XCH	A,@Ri	Exchange indirect RAM with A	1	1
XCHD	A,@Ri	Exchange low-order Digit indirect RAM with A	1	1

BOOLEAN VARIABLE MANIPULATION

Mnemonic		Description	Bytes	Cycles
CLR	C	Clear Carry flag	1	1
CLR	bit	Clear direct bit	2	1
SETB	C	Set Carry flag	1	1
SETB	bit	Set direct bit	2	1
CPL	C	Complement Carry flag	1	1
CPL	bit	Complement direct bit	2	1
ANL	C,bit	AND direct bit to Carry flag	2	2
ANL	C,/bit	AND complement of direct bit to Carry flag	2	2
ORL	C,bit	OR direct bit to Carry flag	2	2
ORL	C,/bit	OR complement of direct bit to Carry flag	2	2
MOV	C,bit	Move direct bit to Carry flag	2	1
MOV	bit,C	Move Carry flag to direct bit	2	2

PROGRAM AND MACHINE CONTROL

Mnemonic		Description	Bytes	Cycles
ACALL	addr11	Absolute subroutine call	2	2

LCALL	addr16	Long subroutine call	3	2
RET		Return from subroutine	1	2
RETI		Return from interrupt	1	2
AJMP	addr11	Absolute Jump	2	2
LJMP	addr16	Long Jump	3	2
SJMP	rel	Short Jump (relative addr)	2	2
JMP	@A+DPTR	Jump indirect relative to DPTR	1	2
JZ	rel	Jump if A is Zero	2	2
JNZ	rel	Jump if A is Not Zero	2	2
JC	rel	Jump if Carry flag is set	2	2
JNC	rel	Jump if No Carry flag	2	2
JB	bit,rel	Jump if direct Bit is set	3	2
JNB	bit,rel	Jump if direct Bit is Not set	3	2
JBC	bit,rel	Jump if direct Bit is set & Clear bit	3	2
CJNE	A,direct,rel	Compare direct to A & Jump if Not Equal	3	2
CJNE	A,#data,rel	Compare immediate to A & Jump if Not Equal	3	2
CJNE	Rn,#data,rel	Compare immed. to reg. & Jump if Not Equal	3	2
CJNE	@Ri,#data,rel	Compare immed. to ind. & Jump if Not Equal	3	2
DJNZ	Rn,rel	Decrement register & Jump if Not Zero	2	2
DJNZ	direct,rel	Decrement direct byte & Jump if Not Zero	3	2
NOP		No operation	1	1

Notes on data addressing modes

Rn	Working register R0-R7
direct	128 internal RAM locations, any I/O port, control or status register
@Ri	Indirect internal RAM location addressed by register R0 or R1
#data	8-bit constant included in instruction
#data	16 16-bit constant included in instruction
bit	128 software flags, any I/O pin, control or status bit

Notes on program addressing modes

addr16	Destination address may be anywhere in 64-kByte program address space
addr11	Destination address will be within same 2-kByte page of program address space as first byte of the following instruction
rel	8-bit offset relative to first byte of following instruction (+127, -128)

APPENDIX C

INTERRUPTS

In order to use any of the interrupts in the MCS-51, the following three steps must be taken.

1. Set the EA (enable all) bit in the IE register to 1.
2. Set the corresponding individual interrupt enable bit in the IE register to 1.
3. Begin the interrupt service routine at the corresponding Vector Address of that interrupt. See Table below.

Interrupt Source	Vector Address
IE0	0003H
TF0	000BH
IE1	0013H
TF1	001BH
R1 & T1	0023H
TF2 & EXF2	002BH

In addition, for external interrupts, pins $\overline{\text{INT0}}$ and $\overline{\text{INT1}}$ (P3.2 and P3.3) must be set to 1, and depending on whether the interrupt is to be level or transition activated, bits IT0 or IT1 in the TCON register may need to be set to 1.

ITx – 0 level activated

ITx – 1 transition activated

IE: INTERRUPT ENABLE REGISTER. BIT ADDRESSABLE.

If the bit is 0, the corresponding interrupt is disabled. If the bit is 1, the corresponding interrupt is enabled.

EA	—	ET2	ES	ET1	EX1	ET0	EX0

EA IE.7 Disables all interrupts. If EA e 0, no interrupt will be acknowledged. If EA e 1, each interrupt source is individually enabled or disabled by setting or clearing its enable bit.

Ð IE.6 Not implemented, reserved for future use.

ET2 IE.5 Enable or disable the Timer 2 overflow or capture interrupts (8052 only).

ES IE.4 Enable or disable the serial port interrupt.

ET1 IE.3 Enable or disable the Timer 1 overflow interrupt.

EX1 IE.2 Enable or disable External Interrupt 1.

ET0 IE.1 Enable or disable the Timer 0 overflow interrupt.

EX0 IE.0 Enable or disable External Interrupt 0.

User software should not write 1s to reserved bits. These bits may be used in future.

This invokes new features. In that case, the reset or inactive value of the new bit will be 0, and its active value will be 1.

ASSIGNING HIGHER PRIORITY TO ONE OR MORE INTERRUPTS

In order to assign higher priority to an interrupt the corresponding bit in the IP register must be set to 1. Remember that while an interrupt service is in progress, it cannot be interrupted by a lower or same level interrupt.

PRIORITY WITHIN LEVEL

Priority within level is only to resolve simultaneous requests of the same priority level.

From high to low, interrupt sources are listed below:

IE0

TF0

IE1

TF1

R1 or T1

TF2 or EXF2

IP: INTERRUPT PRIORITY REGISTER. BIT ADDRESSABLE

If the bit is 0, the corresponding interrupt has a lower priority and if the bit is 1 the corresponding interrupt has a higher priority.

—	—	PT2	PS	PT1	PX1	PT0	PX0

— IP.7 Not implemented, reserved for future use.*

— IP.6 Not implemented, reserved for future use.*

PT2 IP.5 Defines the Timer 2 interrupt priority level (8052 only).

PS IP.4 Defines the Serial Port interrupt priority level.

PT1 IP.3 Defines the Timer 1 interrupt priority level.

PX1 IP.2 Defines External interrupt 1 priority levle.

PT0 IP.1 Defines the Timer 0 interrupt priority level.

PX0 IP.0 Defines the External Interrupt 0 priority level.

User software should not write 1s to reserved bits. These bits may be used in future. This invokes new features. In that case, the reset or inactive value of the new bit will be 0, and its active value will be 1.

TCON: TIMER/COUNTER CONTROL REGISTER. BIT ADDRESSABLE

TF1	TR1	TF0	TR0	IE1	IT1	IE0	IT0

TF1 TCON. 7 Timer 1 overflow flag. Set by hardware when the Timer/Counter 1 overflows. Cleared by hardware as processor vectors to the interrupt service routine.

TR1 TCON. 6 Timer 1 run control bit. Set/cleared by software to turn Timer/Counter 1 ON/OFF.

TF0 TCON. 5 Timer 0 overflow flag. Set by hardware when the Timer/Counter 0 overflows. Cleared by hardware as processor vectors to the service routine.

TR0 TCON. 4 Timer 0 run control bit. Set/cleared by software to turn Timer/Counter 0 ON/OFF.

IE1 TCON. 3 External Interrupt 1 edge flag. Set by hardware when External Interrupt edge is detected.

Cleared by hardware when interrupt is processed.

IT1 TCON. 2 Interrupt 1 type control bit. Set/cleared by software to specify falling edge/low level triggered External Interrupt.

IE0 TCON. 1 External Interrupt 0 edge flag. Set by hardware when External Interrupt edge detected. Cleared by hardware when interrupt is processed.

IT0 TCON. 0 Interrupt 0 type control bit. Set/cleared by software to specify falling edge/low level triggered External Interrupt.

TMOD: TIMER/COUNTER MODE CONTROL REGISTER. NOT BIT ADDRESSABLE

Gate	$C/\overline{T}$	M1	M0	Gate	$C/\overline{T}$	M1	M0
Timer 1				Timer 0			

GATE When TRx (in TCON) is set and GATE = 1, TIMER/COUNTERx will run only while INTx pin is high (hardware control). When GATE = 0, TIMER/COUNTERx will run only while TRx = 1 (software control).

$C/\overline{T}$ Timer or Counter selector. Cleared for Timer operation (input from internal system clock). Set for Counter operation (input from Tx input pin).

M1 Mode selector bit. (NOTE 1)

M0 Mode selector bit. (NOTE 1)

NOTE 1:

M1	M0	Operating Mode	
0	0	0	13-bit Timer (MCS-48 compatable)
0	1	1	16-bit Timer/Counter
1	0	2	8-bit Auto-Reload Timer/Counter
1	1	3	(Timer 0) TL0 is an 8-bit Timer/Counter controlled by the standard Timer 0 control bits. TH0 is an 8-bit Timer and is controlled by Timer 1 control bits.
1	1	3	(Timer 1) Timer/Counter 1 stopped.

TIMER SET-UP

Tables through 6 give some values for TMOD which can be used to set up Timer 0 in different modes. It is assumed that only one timer is being used at a time. If it is desired to run Timers

0 and 1 simultaneously, in any mode, the value in TMOD for Timer 0 must be ORed with the value shown for Timer 1.

For example, if it is desired to run Timer 0 in mode 1 GATE (external control), and Timer 1 in mode 2 COUNTER, then the value that must be loaded into TMOD is 69H (09H from Table ORed with 60H from Table). Moreover, it is assumed that the user, at this point, is not ready to turn the timers on and will do that at a different point in the program by setting bit TRx (in TCON) to 1.

TIMER/COUNTER 0

As a Timer :

Table 3

Mode	Timer 0 Function	TMOD	
		Internal Control (Note 1)	External Control (Note 2)
0	13-bit Timer	00H	08H
1	16-bit Timer	01H	09H
2	8-bit Auto-Reload	02H	0AH
3	two 8-bit Timers	03H	0BH

As a Counter :

Table 4

Mode	Counter 0 Function	TMOD	
		Internal Control (Note 1)	External Control (Note 2)
0	13-bit Timer	04H	0CH
1	16-bit Timer	05H	0DH
2	8-bit Auto-Reload	06H	0EH
3	one 8-bit Counter	07H	0FH

Notes :

1. The Timer is turned ON/OFF by setting/dearing bit TR0 in the software.
2. The Timer is turned ON/OFF by the 1 to 0 transition on $\overline{INT0}$ (P32) when TR0 – 1 (hardware control).

Timer/Counter 1

As a Timer :

Table 5

Mode	Timer 1 Function	TMOD Internal Control (Note 1)	TMOD External Control (Note 2)
0	13-bit Timer	00H	80H
1	16-bit Timer	10H	90H
2	8-bit Auto-Reload	20H	A0H
3	does not run	30H	B0H

As a Counter :

Table 6

Mode	Counter 1 Function	TMOD Internal Control (Note 1)	TMOD External Control (Note 2)
0	13-bit Timer	40H	C0H
1	16-bit Timer	50H	D0H
2	8-bit Auto-Reload	60H	E0H
3	not available	—	—

Notes :

1. The Timer is turned ON/OFF by setting/dearing bit TR1 in the software.
2. The Timer is turned ON/OFF by the 1 to 0 transition on $\overline{\text{INT1}}$ (P3.3) when TR1 – 1 (hardware control).

T2CON: TIMER/COUNTER 2 CONTROL REGISTER. BIT ADDRESSABLE

8052 Only

TF2	EXF2	RCLK	TCLK	EXEN2	TR2	C/$\overline{\text{T2}}$	CP/$\overline{\text{RL2}}$

RF2	T2CON. 7	Timer 2 overflow flag set by hardware and cleared by softwares. TF2 cannot be set when either R CLK = 1 or CLK = 1.
EXF2	T2CON. 6	Timer 2 external flag set when either a capture or reload is caused by a negative transition on T2EX, and EXEN2 = 1. When Timer 2 interrupt is enabled, EXF2 = 1 will cause the CPU to vector to the Timer 2 interrupt routine. EXF2 must be cleared by software.
RCLK	T2CON. 5	Receive clock flag. When set, causes the Serial Port to use Timer 2 overflow pulses for its receive clock in modes 1 & 3. RCLK = 0 causes Timer 1 overflow to be used for the receive clock.
TLCK	T2CON. 4	Transmit clock flag. When set, causes the Serial Port to use Timer 2 overflow pulses for its transmit clock in modes 1 & 3. TCLK = 0 causes Timer 1 overflows to be used for the transmit clock.

EXEN2	T2CON. 3	Timer 2 external enable flag. When set, allows a capture or reload to occur as a result of negative transition on T2EX if Timer 2 is not being used to clock the Serial Port. EXEN2 = 0 causes Timer 2 to ingnore events at T2EX.
TR2	T2CON. 2	Sofware START/STOP control for Timer 2. A logic 1 starts the Timer.
$C/\overline{T2}$	T2CON. 1	Timer or Counter select. 0 – Internal Timer. 1 = External Event Conunter (falling edges triggered).
$CP/\overline{RL2}$	T2CON. 0	Capture/Reload flag. When set, captures will occur on negative transitions at T2EX if EXEN2 = 1. When cleared, Auto-Reloads will occur either with Timer 2 overflows or negative transitions at T2EX when EXEN2 = 1. When either RCLK = 1 or TCLK = 1, this bit is ignored and the Timer is forced to Auto-Reload on Timer 2 overflow.

TIMER/COUNTER 2 SET-UP

Except for the baud rate generator mode, the values given for T2CON do not include the setting of the TR2 bit. Therefore, bit TR2 must be set, separately, to turn the Timer on.

As a Timer :

Table 7

Mode	T2CON	
	Internal Control (Note 1)	External Control (Note 2)
16-bit Auto-Reload	00H	08H
16-bit Capture	01H	09H
BAUD rate generator receive and transmit same baud rate	34H	36H
receive only	24H	26H
transmit only	14H	16H

As a Counter :

Table 8

Mode	TMOD	
	Internal Control (Note 1)	External Control (Note 2)
16-bit Auto-Reload	02H	0AH
16-bit Capture	03H	0BH

Notes :

1. Capture/Reload occurs only on Timer/Counter overflow.
2. Capture/Reload occurs on Timer/Counter overflow and a 1 to 0 transition on T2EX (P1.1) pin except when Timer 2 is used in the baud rate generating mode.

SCON: SERIAL PORT CONTROL REGISTER. BIT ADDRESSABLE

SM0	SM1	SM2	REN	TB8	RB8	TI	RI

SM0 SCON. 7 Serial Port mode specifier. (NOTE 1).

SM1 SCON. 6 Serial Port mode specifier. (NOTE 1).

SM2 SCON. 5 Enables the multiprocessor communication feature in modes 2 & 3. In mode 2 or 3, if SM2 is set to 1 then RI will not be activated if the received 9th data bit (RB8) is 0. In mode 1, if SM2 e 1 then RI will not be activated if a valid stop bit was not received. In mode 0, SM2 should be 0.

REN SCON. 4 Set/Cleared by software to Enable/Disable reception.

TB8 SCON. 3 The 9th bit that will be transmitted in modes 2 & 3. Set/Cleared by software.

RB8 SCON. 2 In modes 2 & 3, is the 9th data bit that was received. In mode 1, if SM2 e 0, RB8 is the stop bit that was received. In mode 0, RB8 is not used.

TI SCON. 1 Transmit interrupt flag. Set by hardware at the end of the 8th bit time in mode 0, or at the beginning of the stop bit in the other modes. Must be cleared by software.

RI SCON. 0 Receive interrupt flag. Set by hardware at the end of the 8th bit time in mode 0, or halfway through the stop bit time in the other modes (except see SM2). Must be cleared by software.

Notes:

SM0	SM1	Mode	Description	Baud Rate
0	0	0	SHIFT REGISTER	Fosc./12
0	1	1	8-Bit UART	Variable
1	0	2	9-Bit UART	Fosc./64 OR Fosc./32
1	1	3	9-Bit UART	Variable

Serial Port Set-Up:

Table 9

Mode	SCON	SM2 Variation
0	10H	Single Processor Environment (SM2 = 0)
1	50H	
2	90H	
3	D0H	
0	NA	Multiprocessor Environment (SM2 = 1)
1	70H	
2	B0H	
3	F0H	

APPENDIX D

INSTRUCTION OPCODES IN HEXADECIMAL ORDER

Hex Number
Mnemonic Operands
Code of Bytes

00 1 NOP
01 2 AJMP code addr
02 3 LJMP code addr
03 1 RR A
04 1 INC A
05 2 INC data addr
06 1 INC @R0
07 1 INC @R1
08 1 INC R0
09 1 INC R1
0A 1 INC R2
0B 1 INC R3
0C 1 INC R4
0D 1 INC R5
0E 1 INC R6
0F 1 INC R7
10 3 JBC bit addr, code addr
11 2 ACALL code addr
12 3 LCALL code addr
13 1 RRC A
14 1 DEC A
15 2 DEC data addr
16 1 DEC @R0
17 1 DEC @R1
18 1 DEC R0

19 1 DEC R1

1A 1 DEC R2

1B 1 DEC R3

1C 1 DEC R4

1D 1 DEC R5

1E 1 DEC R6

1F 1 DEC R7

20 3 JB bit addr, code addr

21 2 AJMP code addr

22 1 RET

23 1 RL A

24 2 ADD A,Ýdata

25 2 ADD A,data addr

26 1 ADD A, @R0

27 1 ADD A, @R1

28 1 ADD A, R0

29 1 ADD A, R1

2A 1 ADD A, R2

2B 1 ADD A, R3

2C 1 ADD A, R4

2D 1 ADD A, R5

2E 1 ADD A, R6

2F 1 ADD A, R7

30 3 JNB bit addr, code addr

31 2 ACALL code addr

32 1 RETI

Hex Number

Mnemonic Operands

Code of Bytes

33 1 RLC A

34 2 ADDC A,Ýdata

35 2 ADDC A,data addr

36 1 ADDC A, @R0

37 1 ADDC A, @R1

38 1 ADDC A, R0

39 1 ADDC A, R1

3A 1 ADDC A, R2

3B 1 ADDC A, R3

3C 1 ADDC A, R4
3D 1 ADDC A, R5
3E 1 ADDC A, R6
3F 1 ADDC A, R7
40 2 JC code addr
41 2 AJMP code addr
42 2 ORL data addr, A
43 3 ORL data addr, Ýdata
44 2 ORL A, Ýdata
45 2 ORL A, data addr
46 1 ORL A, @R0
47 1 ORL A, @R1
48 1 ORL A, R0
49 1 ORL A, R1
4A 1 ORL A, R2
4B 1 ORL A, R3
4C 1 ORL A, R4
4D 1 ORL A, R5
4E 1 ORL A, R6
4F 1 ORL A, R7
50 2 JNC code addr
51 2 ACALL code addr
52 2 ANL data addr, A
53 3 ANL data addr, Ýdata
54 2 ANL A, Ýdata
55 2 ANL A, data addr
56 1 ANL A, @R0
57 1 ANL A, @R1
58 1 ANL A, R0
59 1 ANL A, R1
5A 1 ANL A, R2
5B 1 ANL A, R3
5C 1 ANL A, R4
5D 1 ANL A, R5
5E 1 ANL A, R6
5F 1 ANL A, R7
60 2 JZ code addr
61 2 AJMP code addr

62 2 XRL data addr, A

63 3 XRL data addr, Ýdata

64 2 XRL A, Ýdata

65 2 XRL A, data addr

Hex Number

Mnemonic Operands

Code of Bytes

66 1 XRL A, @R0

67 1 XRL A, @R1

68 1 XRL A, R0

69 1 XRL A, R1

6A 1 XRL A, R2

6B 1 XRL A, R3

6C 1 XRL A, R4

6D 1 XRL A, R5

6E 1 XRL A, R6

6F 1 XRL A, R7

70 2 JNZ code addr

71 2 ACALL code addr

72 2 ORL C, bit addr

73 1 JMP @ AaDPTR

74 2 MOV A, Ýdata

75 3 MOV data addr, Ýdata

76 2 MOV @R0, Ýdata

77 2 MOV @R1, Ýdata

78 2 MOV R0, Ýdata

79 2 MOV R1, Ýdata

7A 2 MOV R2, Ýdata

7B 2 MOV R3, Ýdata

7C 2 MOV R4, Ýdata

7D 2 MOV R5, Ýdata

7E 2 MOV R6, Ýdata

7F 2 MOV R7, Ýdata

80 2 SJMP code addr

81 2 AJMP code addr

82 2 ANL C, bit addr

83 1 MOVC A, @AaPC

84 1 DIV AB

85 3 MOV data addr, data addr

86 2 MOV data addr, @R0

87 2 MOV data addr, @R1

88 2 MOV data addr, R0

89 2 MOV data addr, R1

8A 2 MOV data addr, R2

8B 2 MOV data addr, R3

8C 2 MOV data addr, R4

8D 2 MOV data addr, R5

8E 2 MOV data addr, R6

8F 2 MOV data addr, R7

90 3 MOV DPTR, Ýdata

91 2 ACALL code addr

92 2 MOV bit addr, C

93 1 MOVC A, @AaDPTR

94 2 SUBB A, Ýdata

95 2 SUBB A, data addr

96 1 SUBB A, @R0

97 1 SUBB A, @R1

98 1 SUBB A, R0

Hex Number

Mnemonic Operands

Code of Bytes

99 1 SUBB A, R1

9A 1 SUBB A, R2

9B 1 SUBB A, R3

9C 1 SUBB A, R4

9D 1 SUBB A, R5

9E 1 SUBB A, R6

9F 1 SUBB A, R7

A0 2 ORL C, /bit addr

A1 2 AJMP code addr

A2 2 MOV C, bit addr

A3 1 INC DPTR

A4 1 MUL AB

A5 reserved

A6 2 MOV @R0, data addr

A7 2 MOV @R1, data addr

A8 2 MOV R0, data addr

A9 2 MOV R1, data addr

AA 2 MOV R2, data addr

AB 2 MOV R3, data addr

AC 2 MOV R4, data addr

AD 2 MOV R5, data addr

AE 2 MOV R6, data addr

AF 2 MOV R7, data addr

B0 2 ANL C, /bit addr

B1 2 ACALL code addr

B2 2 CPL bit addr

B3 1 CPL C

B4 3 CJNE A, Ýdata, code addr

B5 3 CJNE A, data addr, code addr

B6 3 CJNE @R0, Ýdata, code addr

B7 3 CJNE @R1, Ýdata, code addr

B8 3 CJNE R0, Ýdata, code addr

B9 3 CJNE R1, Ýdata, code addr

BA 3 CJNE R2, Ýdata, code addr

BB 3 CJNE R3, Ýdata, code addr

BC 3 CJNE R4, Ýdata, code addr

BD 3 CJNE R5, Ýdata, code addr

BE 3 CJNE R6, Ýdata, code addr

BF 3 CJNE R7, Ýdata, code addr

C0 2 PUSH data addr

C1 2 AJMP code addr

C2 2 CLR bit addr

C3 1 CLR C

C4 1 SWAP A

C5 2 XCH A, data addr

C6 1 XCH A, @R0

C7 1 XCH A, @R1

C8 1 XCH A, R0

C9 1 XCH A, R1

CA 1 XCH A, R2

CB 1 XCH A, R3

Hex Number
Mnemonic Operands
Code of Bytes

CC 1 XCH A, R4
CD 1 XCH A, R5
CE 1 XCH A, R6
CF 1 XCH A, R7
D0 2 POP data addr
D1 2 ACALL code addr
D2 2 SETB bit addr
D3 1 SETB C
D4 1 DA A
D5 3 DJNZ data addr, code addr
D6 1 XCHD A, @R0
D7 1 XCHD A, @R1
D8 2 DJNZ R0, code addr
D9 2 DJNZ R1, code addr
DA 2 DJNZ R2, code addr
DB 2 DJNZ R3, code addr
DC 2 DJNZ R4, code addr
DD 2 DJNZ R5, code addr
DE 2 DJNZ R6, code addr
DF 2 DJNZ R7, code addr
E0 1 MOVX A, @DPTR
E1 2 AJMP code addr
E2 1 MOVX A, @R0
E3 1 MOVX A, @R1
E4 1 CLR A
E5 2 MOV A, data addr

Hex Number
Mnemonic Operands
Code of Bytes

E6 1 MOV A, @R0
E7 1 MOV A, @R1
E8 1 MOV A, R0
E9 1 MOV A, R1
EA 1 MOV A, R2

EB 1 MOV A, R3

EC 1 MOV A, R4

ED 1 MOV A, R5

EE 1 MOV A, R6

EF 1 MOV A, R7

F0 1 MOVX @DPTR, A

F1 2 ACALL code addr

F2 1 MOVX @R0, A

F3 1 MOVX @R1, A

F4 1 CPL A

F5 2 MOV data addr, A

F6 1 MOV @R0, A

F7 1 MOV @R1, A

F8 1 MOV R0, A

F9 1 MOV R1, A

FA 1 MOV R2, A

FB 1 MOV R3, A

FC 1 MOV R4, A

FD 1 MOV R5, A

FE 1 MOV R6, A

FF 1 MOV R7, A

BIBLIOGRAPHY

Reference the following books

- The 8051 Microcontroller Architecture, Programming and Applications (Second Edition) – Kenneth J. Ayala.
- Microprocessor and Microcontroller (Second Edition) – R.Theagarajan.
- Microprocessors and Microcontrollers (Third Revised Edition) – A.P.Godse, D.A.Godse.

INDEX

E

F

G

H

I

J

K

L

M

N

O

P